Physics in the Twentieth Century: Selected Essays

Victor F. Weisskopf

Physics in the Twentieth Century: Selected Essays

Victor F. Weisskopf

Foreword by Hans A. Bethe

The MIT Press
Cambridge, Massachusetts, and London, England

This book was designed by The MIT Press Design Department.
It was set in Monophoto Baskerville,
printed by The Colonial Press Inc.,
and bound by The Colonial Press Inc.
in the United States of America.

Library of Congress Cataloging in Publication Data

Weisskopf, Victor Frederick, 1908–
Physics in the twentieth century.

Includes bibliographies.
1. Physics—Addresses, essays, lectures. I. Title.
QC71.W44 530′.08 76–39899
ISBN 0–262–23056–9
ISBN 0–262–73030–8 (pbk.)

To Ellen

Foreword

"Human existence is based upon two pillars: compassion and knowledge. Compassion without knowledge is ineffective; knowledge without compassion is inhuman." Victor Weisskopf said this in a public talk at CERN, Geneva, in 1964, and again in a slightly different form in his essay in this volume, "The Significance of Science."

Viki Weisskopf is a great example of the combination of knowledge and compassion. All his life he has sought and contributed to knowledge, and all his life he has shown compassion. When in the fifties knowledge alone became supreme and any possible application of scientific knowledge to weapons technology had to be pursued regardless of the consequences for the human family, Weisskopf turned to compassion. In this period, when even those of us who disagreed with the popular tendency continued to work on national defense problems while at the same time trying to influence the policy of the United States in what we considered a direction of sanity, Weisskopf appreciated those endeavors but did not join them.

He saw as his main task of compassion the fostering of international collaboration. He worked in France and in England and visited the Soviet Union early and frequently. Most important, he accepted the position of Director General of CERN, the great European Center for Nuclear Research, in the critical period when its principal accelerator, the 30 Gev proton synchrotron, was in its most productive stage. Being an American with European background and not tied to any of the European countries, he seemed most suited to set this great enterprise going. He accomplished its most important goal, namely, to make scientists from the many nations of Western Europe work together in this common task, and, more difficult, to satisfy the governments of all these nations that this cooperation was worthwhile both scientifically and politically. He also encouraged collaboration with Eastern scientists as much as possible.

Having dedicated so much of his life to compassionate endeavor, Weisskopf is most qualified to raise his voice for knowledge now, when so many people call for compassion alone, ignoring or even regarding knowledge as dangerous. His paper "The Significance of Science" is an eloquent appeal to reason. In this essay, he shows how urgently science is needed to solve the problems of the environment that are so much in the foreground of today's interests. The political will to solve them alone cannot accomplish a solution, nor is it enough to use only applied science; fundamental science also is needed, because it alone gives the basic attitude "to tackle any unexplained phenomena and [to be ready] to find new ways to deal with them." He quotes a most impressive list, compiled by Casimir, of the contributions of pure science to useful technology.

He is most eloquent on the intrinsic value of science, "Science is a truly universal human enterprise; the same questions are asked by all men involved in science; the same joy of insight is experienced when a new aspect of deeper coherence is found in the fabric of nature." And "Basic science leads to an intimate relation between man and nature, to a closer contact with and a deeper understanding of the phenomena."

A most beautiful and concise statement of the scientists' creed is in Weisskopf's "Science and Ethics," presented at a conference on the Philosophy of Science. He admits that science has shaken the supernatural system of thoughts. But he points out, against this, that it has established the validity of natural laws and given insight into the order of the world: man is not at the mercy of a capricious universe. While popular belief accuses science of having made everything relative, Weisskopf points out that relativity theory is really a theory of the absolute, of the quantities that remain unchanged from any standpoint (invariants). Quantum theory is similarly misunderstood. Scientific theories are not overthrown; they are expanded, refined, and generalized.

Another accusation against science is that it represents nature as a mathematical formula and, thereby, has dehumanized our relations with the world around us. On the contrary, scientific endeavor has so many cooperative aspects that it makes the scientific community "supranational because it transcends national and political differences." This common attitude is reflected in a certain state of mind within the scientific community and "is opposed to some of the negative and destructive trends in today's thinking."

My friendship with Viki Weisskopf started in 1937 when he became instructor in physics at the University of Rochester. Nearly every month we visited each other, discussing mainly questions of physics but also the great political upheaval that culminated in the Second World War and many other human and personal problems. In physics we were both interested in the theory of the atomic nucleus that was then in its first great blossoming.

Before coming to the United States, Weisskopf had gone through the most excellent schools of theoretical physics in Europe. Born and raised in Vienna, he took his Ph.D. in Göttingen with E. P. Wigner in 1931. In 1932 to 1933 he obtained a Rockefeller Foundation Fellowship and went to Copenhagen and to Cambridge, England. The next three years he spent in Zürich with W. Pauli, the man who had probably the deepest understanding of quantum mechanics. The following year he returned to Copenhagen to be again with Niels Bohr, the great master of twentieth century physics. Weisskopf's beautiful appraisal of Niels Bohr's work and life is included in this collection ("Niels Bohr, the Quantum and the World"). From Copenhagen he emigrated to the United States and spent the next five and one-half years in Rochester.

In 1943 I persuaded Weisskopf to join the Los Alamos Scientific Laboratory to help with the development of the atomic bomb. He was my deputy as leader of the theory division of the laboratory and was also a group leader. He was the division's most popular member with the experimental physicists, who came to him constantly to ask his opinion about the magnitude of this or that cross section to determine whether their next experiment had to be an elaborate or could be a simple one. His opinions were nearly always correct, and after a while somebody put a sign in his corridor pointing the way to the "Los Alamos Oracle."

His warmth and compassion were felt keenly at Los Alamos: He was elected Chairman of the Town Council, an informal body which dealt with the many human problems that occurred in this otherwise ungoverned community.

In 1946 Weisskopf joined the faculty of MIT, and has stayed there ever since, except for his excursions to Europe and to CERN in particular. He has become an elder statesman, much loved and admired by his colleagues, whose advice is sought, respected, and generously given. Among other things he is at present chairman of High Energy Physics Advisory Panel (HEPAP), a scientific committee giving policy advice on the construction and operation of the high energy accelerators in the United States.

Weisskopf has written a large number of important scientific papers on field theory and electrodynamics, on nuclear physics, and on high-energy and particle physics. Three of his contributions changed the course of physics: The first was a theory of the width of atomic energy levels and its fundamental relation to their lifetime. The second, done at Rochester, showed that the self-energy of an electron diverges only logarithmically in the Dirac theory, a discovery that greatly helped in the development of the theory of renormalization in 1947 to 1948 (see below). The third is known as the "cloudy crystal ball" in which he showed that nuclear particles continue to have an individual existence in the nucleus in scattering problems. This contradicted Niels

Bohr's theory of the compound nucleus, but it explained some striking and puzzling experiments. It has been accepted as one of the fundamental features of nuclear physics.

Weisskopf's writings fall into four groups. First, there are the purely scientific papers, including his textbook (with Blatt) on nuclear physics. Second, he has written summary articles on special topics. As stated in the introduction to "The Quantum Ladder," "no meeting of nuclear physicists is quite complete without a summation by Viki." Third, there is a series of articles on the main themes of physics, interpreting the quantum theory and leading to high-energy physics. Finally, there are the already discussed writings on science in general which might best be called "natural philosophy."

His summary articles on special topics are lucid and are "popular scientific writing" at its best. The subject is made clear to physicists in other branches of physics and to the educated public, by simplifying it, but without ever making a compromise with fundamental accuracy, as many popular writers on science do when they get enamored with their own imagery and forget the subject that they really want to explain. He is known as an exquisite interpreter of science from his delightful book, *Knowledge and Wonder.*

The summary article about electron theory ("Recent Developments in the Theory of the Electron," 1949) is written essentially for physicists. He summarizes the great advances brought about by renormalization theory which gives finite results for all observable quantities and is fundamental to modern field theory. His own field of nuclear physics is treated in the essay "The Compound Nucleus"; it is summarized in "Problems of Nuclear Structure" (1961) and in a simpler form in "Nuclear Models" (1951). Weisskopf is always excited by important developments in physics even if he himself has not contributed. A major one of these was the "Fall of Parity." This article is a beautiful example of his ability to make things understandable without changing the essentials. In "The Visual Appearance of Rapidly Moving Objects" he explains the discovery of Penrose that a rapidly moving object looks to us as if it had its natural shape and is not distorted by the Lorentz contraction as had previously been believed; in fact without relativity theory it would look badly distorted. His "How Light Interacts with Matter" is rich in new insights.

Viki's main concern, as with most theoretical physicists, is the correct interpretation of the quantum theory. In many variations, he explains the fundamental achievement of the quantum theory: that it makes us understand why all carbon atoms are exactly alike, as are all the atoms of every chemical element. This fact is fundamental to the world around us and can in no way be explained in terms of "classical" physics. The uncertainty principle, which is commonly quoted as the essence of quantum mechanics and which seems to dissolve atomic mechanics into a haze of probabilities, is only incidental, although it is of course necessary to avoid paradoxes. So rather than making our world more uncertain, quantum mechanics makes it more definite; it is the cause of the dependability of the world to which we are accustomed.

This theme is already prominent in 1959 in his article "Quality and Quantity in Quantum Physics." Here he makes clear how a quantative difference, such as the number of electrons of 6 in a carbon and 11 in a sodium atom, makes all the difference in the qualitative behavior of these two chemical elements. He also foresees, at this early time, some of the developments in biology related to DNA.

The theme is further developed in "Physics in the Twentieth Century" and "The Quantum Ladder." Here, and especially in "Quantum Theory and Elementary Particles," he also emphasizes the essential similarity of atoms and nuclei, and then of fundamental particles. I don't know any other place where the essential solidity of our world as governed by quantum theory is better explained, or the great richness of the phenomena caused by quantitative differences is shown with greater love.

"The Quantum Ladder" is natural philosophy. Viki expands this philosophy to subjects outside physics in such papers as "Man and Nature." Here he says that the manifold arrangements of organic molecules are a consequence of quantum theory. Reproduction of the same molecule is again a consequence of some of these intricate arrangements, as is biological evolution. This evolution goes on a time scale of millions of years. The nervous system of the higher animals, and especially of man, has made it possible to have much more rapid evolution, based on the accumulation of knowledge and the transmission of this knowledge from one generation to the next. This social evolution of man has been measured in thousands of years. Another step was then added by science in which man learned to search for the causes of natural phenomena and to use this understanding to create his own environment: This has now compressed the time scale of evolution to about one human generation.

I hope the reader will find as much pleasure as I did in reading the works of Victor Weisskopf from many fields, of Viki the scientist and of Viki the humanist.

Hans A. Bethe

My Life as a Physicist

I was born in 1908 in Vienna, Austria. I spent my childhood under the Austrian-Hungarian monarchy and lived through the First World War as a child. I witnessed as a child the collapse of the monarchy and went through my secondary education in a time of great changes. I owe much to the cultural traditions of Vienna from Mozart and Beethoven to Freud and Boltzmann. I entered Vienna University and started to study physics in the fall of 1926. It was just a few years too late to have been an active participant in the creation of early quantum mechanics. Many of my good friends were a little older, men such as Hans Bethe, Felix Bloch, Rudolf Peierls, or Walter Heitler. Three years would have made a big difference at that time. I came to the university in 1926 after quantum mechanics was invented, and, of course, I needed a few years to learn physics. That meant I could not start active work in physics before 1929–1930, and all the fundamental developments in quantum mechanics were made between 1925 and 1930. I made the interesting observation that three or four years mean a lot when you are very young; they mean nothing during your mature life; and they again mean a lot when you are getting old. Today, I am not unhappy about being three years younger!

I was conscious of the existence of modern physics before I went to the university. I read popular books on physics when I was in the gymnasium; at that time a lot of popular literature was written about the Bohr atom and the quantum theory. In my youth there was a very good science periodical named *Kosmos*. There were quite a few articles about the Bohr atom; I remember a semipopular description of the explanation of the periodic table by Niels Bohr by means of the *Aufbauprinzip* that made a terrific impression on me. It was a fantastic revelation of how simply one can explain all these regularities.

Lecture given at the Erice Summer School in High-Energy Physics, Sicily, Italy, in 1971.

But this was still on a semi popular level, and my first real contact with quantum mechanics came later when I actually started my studies. Of course, at that time quantum mechanics was not taught to freshmen. (Maybe it shouldn't be now either, but it is.) At that time certainly it was not, and I got in touch with quantum mechanics roughly in 1928 when I came to the University of Göttingen. The previous two years, which correspond roughly to undergraduate studies, were spent in Vienna; there I learned physics from Hans Thirring, who was a good physicist and an excellent pedagogue. He gave a series of lectures on theoretical classical physics (which to my mind belong to the best one can get in physics education), mechanics, optics, electrodynamics, thermodynamics. It was great! After two years, I asked Thirring, "What shall I study now?" He said, "One thing I must advise you: If you want to learn more physics, go away from Vienna."

I took his advice and went to Göttingen. At that time a student had two choices of outstanding schools in physics in Germany: He could go to Göttingen, where there was Max Born, James Frank, P. Jordan, W. Heitler, and many others. Or he could go to the other center in German-speaking universities, Münich, where Sommerfeld was and his great school. I think I made a mistake in going to Göttingen. If I had gone to Münich I would have received a much better training in mathematical physics, a training I badly needed and missed all through my scientific life.

Anyway, I did go to Göttingen, and there I met the great physicists who were the creators of quantum mechanics. However, it was not as important to meet the great ones as the many young people who were on my level or a little more advanced, all eager to learn, to discuss, and to argue. But there was one older physicist who made a greater impression on me than any of the others and shaped my attitudes toward physics: He was

P. Ehrenfest. I think he was in Göttingen only one semester in 1929, in order to replace Max Born, who was sick. Ehrenfest took a personal interest in me—we both were born in Vienna—and taught me for the first time to distrust the complicated mathematics and formalisms that were then very popular in Göttingen. He loved to ask, and encouraged others to ask, "stupid" questions; he refused to admit that something is understood if one understands only the mathematical derivation. He showed me how to get at the real physics, how to distinguish between physics and formalism, how to get at the depth of things: "Physics is simple, but subtle," he used to say. The older I get, the more aware I am of his influence.

Let me return to the young people then in Göttingen. There were young instructors such as Heitler, Herzberg, Nordheim, and Wigner, or graduate students of my age or a little older, such as Delbrück and Teller and Maria Mayer. We learned quantum mechanics together and discussed it with each other. It was a great experience and sensation. I will never forget G. Herzberg—he also remains in my memory as one of the outstanding teachers—he at that time taught for the first time the course that is now immortalized in his book which you all know, *Introduction to Atomic Physics*. It was an outstanding course at that time and still is, a fantastic achievement after forty years! Atomic quantum mechanics, the consequences of symmetry, the term spectra, the systematics of spectroscopy, the transition probabilities, all this was exciting because it was so beautiful, and only a few years old.

It was a great thing and very disagreeable too to be a graduate student at that time. It was disagreeable because too many new ideas came around. When you had barely digested the Schrödinger equation and Heisenberg's quantum mechanics, you already heard your colleagues talking about the Dirac equation and quantum electrodynamics. Learning went a little too fast; it was interesting, but discouraging. Let me return to my statement that I came three years too late. Those fellows such as Bethe, Peierls, Bloch, and Heitler were lucky. Every Ph.D. thesis at that time opened up a new field. Peierls worked on heat conduction and opened one part of solid-state physics. Bethe wrote his Ph.D. paper on electron diffraction of crystals and opened up another part of solid-state physics. Heitler and London opened up quantum chemistry, Wentzel the theory of the photoeffect.

It was already too late when I entered; all the fields had been opened. What did I do? When I came to my doctoral thesis I was thinking about quantum electrodynamics. Dirac had published his first paper on quantum electrodynamics, on the interaction of light with particles. I was especially interested in the question of radiation damping, the natural width of spectral lines. I dabbled around alone and tried to find exponential solutions of electrodynamics; I did not get very far because I was too young and ignorant. I asked the great Wigner for help; he was a few years older than I, but he already was very famous at that time. Of course, he helped me, right away; together we wrote a paper on the natural width of spectral lines, a paper that contains for the first time a divergent integral. I tried to convince Wigner that the integral could be made to vanish. Wigner said, "No, no, it is infinite." I didn't believe him, but he was right, of course. This paper, part of which later became my thesis, was the first paper in which divergent integrals appeared. They have not yet been resolved; they are still there after 40 years. One ought to be ashamed of it. Another remark about this paper: Imagine a young man completely unknown, whose name starts with the letter W and has the luck to publish his first paper with Wigner, so that the paper is Weisskopf-Wigner and not Wigner-Weisskopf? At that time I made a vow and said

"The gods were favorable to me by letting me work with Wigner, so I shall from now on, whatever paper I write and with whatever man, I shall always stick to the alphabetical order of authorship." And I have done so all my life with one exception, when a student of mine changed it without my knowledge. I had to fight for it when I wrote a book with someone you may know, namely, Blatt and Weisskopf. The publishers insisted that it should be Weisskopf and Blatt, because I was supposed to be the senior author. I said, "I cannot do this; I have made a vow and I cannot break a vow." They said, "We don't care about vows, we want you first, you are the senior author, sir," and I finally convinced the publishers only by pointing out that if it is "Weisskopf and Blatt," the emphasis is on Blatt, but if it is "Blatt and Weisskopf" the emphasis is on me, and they accepted it. So much for that.

Another more interesting story from this period: It was a very exciting time, because of the invention of the Dirac equation. Everybody was completely flabbergasted that Dirac, out of sheer intuition, could write down an equation that completely explained the properties of the relativistic electron. In fact, the equation had much more in it than even Dirac thought: It contained the positron also. But that came later. I was in Göttingen 1928 to 1931; the Dirac equation was published in 1927, and everybody discussed the Dirac equation and the fact that the gyromagnetic factor of the electron turns out to be two. How wonderful!

A good example of the arrogance of theorists, which has not changed in 40 years, is the following story: There was a seminar held by the theoretical group in Göttingen, and Otto Stern came down to Göttingen from Hamburg and gave a talk on the measurements he was about to finish of the magnetic moment of the proton. He explained his apparatus, but he did not tell us the result. He took a piece of paper and went to each of us saying,

"What is your prediction of the magnetic moment of the proton?" Every theoretical physicist from Max Born down to Victor Weisskopf said, "Well, of course, the great thing about the Dirac equation is that it predicts a magnetic moment of one Bohr magneton for a particle of spin one-half." Then he asked us to write down the predictions; everybody wrote "one magneton." Then, two months later, he came to give again a talk about the finished experiment, which showed that the value was 2.8. He projected the paper with our predictions on the screen. It was a sobering experience.

Let me now say a few words which have some bearing on the situation in the 1970s. I am not sure whether you know that in the history of the economic development of the Western world 1931 was a very bad time, even worse than 1971, especially for young physicists looking for jobs. It was a most terrible time both economically and politically. Hitler was already quite powerful; in fact, everybody predicted that Hitler would take over Germany. Also, there was a disastrous economic depression, both in America and in Europe. So, what should young physicists do? Of course, positions were not available. The situation was much worse than now; even physicists such as Bethe and Bloch had a difficult time getting jobs. It was very difficult even for the best. There was a difference: Then we had no expectations. Today students are deluded: If they studied physics, they would find jobs; they would have ten jobs offered; and they need not be so very good to get well-paid jobs. Forty years ago, we never had such expectations. In fact, my father told me that I was crazy to study physics and that I should study something that was more realistic, such as engineering or another "practical" profession. Because we did not grow up with this expectation, it was easier to live with temporary half-jobs.

Anyhow, I did not have a job. First, I went to Heisenberg as an unpaid "post-doc," because paid post-docs did not exist at that time. I asked my parents for a little money and had a very good time with Heisenberg and his collaborators. That year, 1932, was a fantastic year in physics: Fermi published his theory of weak interactions; the neutron was discovered by Chadwick; and Anderson and Neddermeyer discovered the positron. The base was laid of weak interaction, of strong interaction, and of the concept of the anti-world. For us, the appearance of another paper by Fermi, his simple formulation of quantum electrodynamics, was a revelation. After having ploughed through longish papers by Jordan and Wigner, and Pauli and Heisenberg, Fermi's formulation was so simple and so clear that even a nonmathematically inclined physicist could understand it. We had an interesting time in Leipzig. I would like to relate a story that happened there.

I remember discussions in our group why the electron and the neutrino come out of the nucleus. Nowadays, of course, every student knows that the pair is created, but at that time the idea of particle creation was quite new; the positron theory was not yet conceived. I remember we were sitting in a coffee house in Leipzig that faced the entrance to a swimming pool; Heisenberg looked at the door. We were discussing how is it possible that an electron comes from the nucleus without having been there before. He said, "You are all making the wrong conclusions; look at that door. You see everybody entering fully dressed and coming out fully dressed; would you conclude from this that the swimmers inside are fully dressed?"

A few words about my job problems: Because F. London, who was the permanent assistant to Schrödinger, was asked to go to America for half a year, I was lucky enough to get a job for half a year in Berlin as an assistant to Schrödinger. Then, afterward when the money from home became scarce, I decided I had to go somewhere where I could support myself. I would not get a job in Germany, or England, or France. In 1933 I went to

Kharkov, Russia for almost a year, where it was possible to get a job with subsistence. Working in Kharkov at that time were Landau, Lifschitz, and Achiezer, and many other young Russian physicists. Life in Russia was by no means easy. There was a serious famine because of the difficulties of collectivization. But life was interesting and instructive.

Later, I had a stroke of luck (I think it was Schrödinger who did this for me): One day I got a letter from America telling me that I had received a Rockefeller fellowship. (They are no longer given to physicists today.) That fellowship meant one year's support anywhere; the only requirement was to report what you had done. I do not know whether you can imagine what this meant when I was tempted to give up hope and thought I would have to give up physics. (I did teach in high school for a time and enjoyed it.) The Rockefeller grant was impressive; at that time prices were very different, the $1800 a year was unexpected wealth. I decided to use the grant to go to Copenhagen to the great Bohr, and afterward, for the second part of the year, I would go to Cambridge, to the great Dirac.

When I arrived in Copenhagen, I was met by my friend, Max Delbrück, who is now a biologist, but at that time was a physicist. He told me that Copenhagen was a wonderful place, mainly because of its beautiful girls. I said, "That is very interesting information, but it has no value for me because I would like to do some physics." Delbrück said, "Well, yes, but at least you could have a look." The next evening he took me to some dancing place with some of his friends, and indeed, the information turned out to be correct: One of them is now my wife. A wife is of great importance in the life of a physicist; she makes life bearable to him if he makes it bearable to her. There were other attractions in Copenhagen, such as Niels Bohr and his institute. There was an interesting group there at that time: Felix Bloch,

G. Placzek, M. Delbrück, Chandrasekhar, and J. E. Williams, the famous Williams of Weizsacker and Williams, and Weizsacker also. At that time I was still interested in quantum electrodynamics, and there were two problems to be faced: One was the problem of the positron, whether it really is contained within the Dirac equation, the problem of the charge conjugation symmetry, as we say today, and two, the problem of the nuclear force, the beginning of nuclear physics. Here was something completely new; it was like a bombshell: A new force was discovered in nature, what should we do with this nuclear force? Could we apply quantum mechanics again? All discussions focused around nuclear structure on the one hand and quantum electrodynamics on the other. I stayed in Copenhagen for more than half a year because of all these attractions; however, in time I thought I should go to Dirac after all, and I moved for a few months to Cambridge, England.

In Cambridge I was a little disappointed; Dirac was a very great man, but he was absolutely useless for any student. You could not talk to him, or, if you talked to him, he would just listen and say, "Yes." From a student's point of view, Dirac was a lost experience. There I met Peierls, who also had a Rockefeller fellowship. I learned an enormous amount from him. He introduced me to the complicated calculations of electrodynamics, the so-called alpha-gymnastics.

Then again I had another stroke of luck. I got a letter from Wolfgang Pauli, telling me that I should come to him as an assistant. This was Pauli, the great Pauli. I did not know him at that time; I had only heard about him. I had seen him at some of the conferences, of course, and I was tremendously impressed, first, that I got a job—the job situation was still very bad—and second, a job with Pauli. I went to Peierls and said "I just got a job from Pauli!" Peierls had been an assistant to Pauli a few years before and told me "If you think that is a desirable job, you are wrong; it is a terrible thing to be assistant to Pauli." I asked why. "You will see!" He gave me a few good words of advice that I will talk of later. In a few weeks, Pauli asked me to come to Zürich. I came to the big door of his office, I knocked, and no answer. I knocked again, and no answer. After about five minutes he said, rather roughly, "Who is it? Come in!" I opened the door, and here was Pauli—it was a very big office—at the other side of the room, at his desk, writing and writing. He said, "Who is this? First I must finish calculating. *Erst, muss ich ixen.*" Again he let me wait for about five minutes and then: "Who is that?" "I am Weisskopf." "Uhh, Weisskopf, ja, you are my new assistant." Then he looked at me and said, "Now, you see I wanted to take Bethe, but Bethe works now on the solid state. *Den Festen Korper mag ich nicht.* Solid state I don't like, although I started it. This is why I took you." Then I said, "What can I do for you, sir?" and he said, "I shall give you right away a problem." He gave me a problem, some calculation, and then he said, *"Gehen sie,* go and work." So I went, and after 10 days or so, he came and said, "Well, show me what you have done." And I showed him. He looked at it and exclaimed: "I should have taken Bethe!" He was probably right. Anyhow, I enjoyed the work no end. Peierls gave me good advice, he couldn't have predicted this story, but he predicted many other things. For example, Pauli was always very disagreeable during a colloquium; he was dramatic, I mean. At one colloquium Pauli was sitting in the first row and when he didn't like a statement he would get up and say: "This is wrong! *Falsch!*" "Go home, you are not prepared, such nonsense, how can you tell such nonsense!" He would tell that to anybody, big shots or small shots. Now Peierls gave me the following advice: "If you ever give a colloquium, the following procedure should be followed: If the colloquium is to be given in the afternoon, that morning you go to Pauli and say 'Professor Pauli, I want to tell you what I am going to say this afternoon' and then you tell him. He will start scolding you, saying 'Nonsense, *Dummheit* and go—go home.' You take it all, you don't need even to listen to it, and then, in the afternoon, you give your colloquium exactly as you prepared it. Nothing will happen, because Pauli will sit in the first row and he will listen and mutter to himself 'Well, I have told him already!' " It worked every time.

I loved working with Pauli. First of all, my admiration had no limits. (We really need Pauli today; we need his wonderful criticism, his positive attitude, and his no-nonsense attitude against anything that is not really good.) I learned a great deal of physics from him, and I enjoyed it tremendously.

One of our joint works was a paper that turned out to be the beginning of meson theory, the quantization of the scalar wave equation. In the beginning, I had some vague ideas about the Klein-Gordon equation; in particular, I was struck by the fact that the total number of particles, interpreted as the absolute square of the wave function, was not conserved. I thought that maybe there was something like pair creation there, but it didn't work well, because I was very inept in applying second quantization. So I tried to tell Pauli the problem. He was in bad humor. (In spite of his occasional rough manners, he was one of the kindest and warmest persons I ever met.) I tried to tell him, "Listen, Pauli, there is something which I think is interesting" and then he said, "*Dummheit*, nonsense, go away." Again and again I tried, "Look, tell me, I want to ask a question." "*Dumm, dumm.*" Then I became angry and told him a quotation "*Ach, Meister, warum so viel Eifer warum so wenig Ruh, mich duenkt euer Urteil waere reifer, hoertet ihr besser zu.*" This means, in English, "Oh, master, why so much excitement and so little calm; I think that your judgement would be more mature if you listened more carefully." He then looked at me and said, "What is that?" and I said, "That is from Richard Wagner, *Die Meister-*

singer." And then he said, "Wagner? *Mag ich ueberhaupt nicht!*" "Wagner? I don't like him at all." And then, of course, it was over, I had to go. Two days later, I again said, "Look, there is an interesting problem here," and he said "Uh, why didn't you tell me this right away?" Then a wonderful collaboration began, and we had tremendous fun in working out something that, at that time, was quite unexpected: that one can get pair creation and pair annihilation without a Dirac equation, also for particles without spin.

At that time I was also interested in the study of the electromagnetic properties of the vacuum, the polarization of the vacuum by electric fields, and pair creation. I collaborated on these topics with Euler and Kockel, two students of Heisenberg. Euler, an excellent physicist, the grandson of the famous mathematician, was later killed by the Nazis. We were interested in the infinities occuring in electrodynamics; if we had been only more clever, we could have done the whole renormalization theory at that time: it was 1934. We had it, the main ideas were formulated in a paper that I wrote in 1934. We could show then that the divergent expressions are essentially unobservable—in other words, that they can be incorporated into a renormalized mass and charge. So, the whole idea of renormalization was there, but we did not sit down and work it out. We did it in a very primitive way.

In those days I derived—on a suggestion by Pauli—the electromagnetic self-energy of the electron, and I could show how positron theory decreases the degree of divergence to a logarithmic one. I must admit that I made a bad mistake in my first publication; Wendell Furry pointed it out to me, and the correct result emerged in a second publication based on Furry's remark.

Later on, nuclear physics became the center of interest, first, because an increasing amount of experimental results became available and, second, because electrodynamics rather became stuck. In 1935, my work shifted to nuclear physics, especially after Bohr invented the compound nucleus picture. I worked on the concept on nuclear temperature and applied thermodynamics to small units; it was interesting, and I learned a great deal about nuclei and about thermodynamics. Then I had to leave Pauli and Zürich and got a sort of auxiliary position with Niels Bohr. It was paid from funds designated to support scientists who were out of jobs because of the Hitler persecution. Such jobs were, of course, only temporary, but it was good to be in Copenhagen again. Bohr went to America for one of his regular visits, where he tried to get jobs for his refugee physicists, who accumulated in Copenhagen. One day, in 1937, he came back and said he had a position ready for me at Rochester, New York. It was an instructor position for $2,400 a year. I was glad to part from a Europe that did not look very promising at that time.

Rochester was a very pleasant little university—and still is, both pleasant and little. It had a very good physics department at that time (it still has) with a good deal of nuclear physics; they had a cyclotron and did quite advanced work, nuclear excitations and penetrations of potential barriers. My experience in nuclear physics turned out to be very useful. We had an extremely pleasant life there far away from the Nazi horrors and the uncertainties of life in a war-threatened Europe. We met a lot of physicists, both American and European. There were many recent immigrants, for example, Bethe, who was nearby in Ithaca, only a few miles from Rochester. In the summers we went to the West Coast, visiting Berkeley and Stanford where R. Oppenheimer and F. Bloch had assembled a large and active group of physicists.

In thinking back on those first years in the United States, I cannot help being deeply impressed by the way we were received by the American community of scientists. We were warmly welcomed, treated as equals and sometimes even as superiors; immediately we felt at home and part of the community, as if we had been there from the beginning. In no other country—and we lived in many—did we feel so quickly accepted and appreciated.

World War II broke out in 1939, two years after we arrived in America. First America stayed out of it, but science and the scientists were deeply affected long before America openly entered the war in 1941. Most physicists either went to work on radar or on the atomic bomb. The atomic bomb was a later development, so most of my friends, Americans and Europeans, worked on radar. J. Schwinger was the great radar expert, and he developed the theory of wave guides at that time. Some of the Europeans, who like myself, had come relatively recently to the United States were still considered as "enemy aliens" coming from Germany or Austria and were not admitted to war work but were asked to take over the teaching of students. At that time I taught theoretical physics at Rochester, at Ithaca, and at other places, in order to replace scientists who had gone to the war laboratories. It was a rather interesting experience.

Let me now return to the discovery of fission and the development of its use for a bomb and for power production. In 1942 Fermi succeeded in setting up a nuclear chain reaction and the technical applications of the fission process became a realistic possibility. Many physicists were asked to join in the effort of development, whether they were "enemy aliens" or not. The brainpower was needed. At the beginning of 1943 Oppenheimer asked me to join a group of physicists at Los Alamos to develop the atomic bomb.

It is very difficult to describe—let alone to judge—one's feelings and one's decisions after 30 years. There are many now who say that a physicist should devote his time to the search of truth and should not devote his abilities to produce weapons of destruction. It is easy to say, and it may even be right. There were very few who actually followed it. Pauli had to leave Europe at that time and was in Princeton and he never even thought of joining any of these war projects. But for a man in my position, and for so many others, not to join would have been a very difficult path to follow. I cannot tell you why I went to Los Alamos. The fear that Hitler would make the bomb had much to do with it. If someone suggests I went just because everybody else did, he may be right. The significance of physics is a very difficult thing to describe. Physics is not only the search of truth, it is also potential power over nature; the two aspects cannot be separated. One may say that a physicist should search only for truth and should leave the power over nature to somebody else; but this attitude actually skirts the subject and does not look at reality. The important point about physics—and of most of science—is not only that it represents natural philosophy but that it also is deeply involved in action—in life, in death, in tragedy, in abuse, in the human predicament. Whether this is good or bad, who is to judge? Both sides can be argued. Some terrible things have happened since man made use of fission. However, the first bomb may have ended the war. At least, that is what we thought; we thought we had saved a million lives; maybe we were right. The second bomb was much harder to defend. We did participate in all that and I don't want to condemn or to justify it.

I cannot deny that those four years in Los Alamos were a great experience, from a human as well as from a scientific point of view, in spite of the fact that they were devoted to the development of the most murderous device ever created by man. Such are the contradictions of life. From the human point of view, to live together, intellectually and otherwise, with the best physicists from the whole world (such as Niels Bohr, Enrico Fermi, J. Chadwick, R. Peierls, E. Sègre, and many more) was quite an experience. The discussions we had on philosophy, art, politics, physics, and on the future world under the shadow of a super-weapon remain unforgettable. But also from the purely professional level, we had to face tasks that never were faced before. It was a remarkable experience to deal with matter under unusual conditions. We attempted to predict the behavior of matter and make experiments about it under conditions that were by factors of thousands outside of the ordinary. To hold in your hand a piece of metal that man made all new, to see events and processes actually develop before your eyes which never had been seen before, that was something very remarkable.

Then came the great test, the first nuclear explosion. I was in charge of theory at the test site in a New Mexico desert that had the interesting name "Jornada del Muerte," which means something like "journey of death." I had to go from one station to another in a jeep and to tell them the expected intensities of this or that. Of course, we had calculated what the safe distance for a man would be. They said they wanted to put me in an iron cage at exactly that distance the theorists proclaimed safe. We were wrong by a factor five; I would have been dead at the theoretically safe place. The gamma ray intensity was much higher than we thought, and we misjudged the energies and the absorption.

At the end of the war we were all glad to go back to our regular lives of teaching and research. There was hope for a peaceful era, for a new and better world than what was left in shambles after the most murderous and destructive conflict in history. We felt again the immense desire to apply our minds to basic research. We wanted to make use of the many new tools and methods that emerged from the wartime developments, but for the purpose of finding out more about nature instead of for developing more efficient weapons. We took up with great enthusiasm our old as well as many new problems, and we were amply supported by the government authorities who had become aware of the power of physics.

After the war I became Professor of Physics at the Massachusetts Institute of Technology. Let me tell you an experience I had at that time: There were vague rumors since 1936 that the observed hydrogen levels did not exactly coincide with the predictions of the Dirac equation—the so-called Pasternak effect. There were some speculations as to how this effect could be calculated with our divergent quantum electrodynamics. After the war, I decided to take this problem up with a very good graduate student, Bruce French, who is now well known in nuclear structure physics. We thought of calculating this effect, commonly known as the Lamb shift, by trying to isolate the infinite self-energy of the electron. It was a difficult calculation, because the renormalization techniques were not developed yet; we had to calculate the difference of the energy of the free electron and of the bound electron, both being infinite. We had to be very careful because taking differences of divergent magnitudes is risky. But we slowly worked our way through the difficulties, since there were no really good experimental results available. Then one day, Lamb and Rutherford made a good experiment, and we worked a little harder, but a lot of other people also did so. We finally got a result that fitted the experiments quite well. I told our result to Julian Schwinger and Dick Feynman. Schwinger was across the street at Harvard and Feynman was at Cornell at that time; they repeated our calculations but did not get our number. Schwinger got the same number as Feynman, but it was different from ours. I said to French: "Well, the probability is high that they are right and that we are wrong." We postponed our publication in order to find the mistake and searched for half a year. In the meantime Lamb and Kroll published their calculations of the same effect which were more or less identical with ours. I then got a telephone call from Feynman in Ithaca, "You are right, I am wrong!" Now, if we had the guts to publish our own results, our paper would have been the first one explaining the experiments of Lamb and Rutherford. What was the lesson of this story? You've got to believe in what you are doing; you get Nobel prizes not for what you do but for perseverence. Our fault and our weakness was that we didn't have the necessary belief. Of course, later on, the methods we used were refined and improved, and today every student can do it in a few hours, but that is the progress of—I wouldn't say of science—but of the mathematical formalism. The two are not the same.

For quite some time after this, my main interest turned to nuclear physics. My collaborators, mainly Herman Feshbach, and I developed some new ways of looking at and explaining nuclear resonances and scattering effects. One of our models described the nucleus as a transparent and partially absorbing sphere. It is often referred to as the "cloudy-crystal-ball" model, but it did reproduce many properties of the nucleus and did so rather well. I wrote a fat book on *Theoretical Nuclear Physics* with John Blatt, and both of us learned a great deal of physics in doing so. I discovered that the effort of explaining and clarifying a field of physics not only leads to a better understanding of past work but also produces many new ideas, explanations, and discoveries. This is yet another illustration of the close connection between teaching and research. It never seemed possible to me to do one without the other. At the time among the problems in which my collaborators and I were most interested was the surprising validity of the shell model of the nucleus. In this model the nuclear constituents are seemingly moving as almost free particles within the nucleus, in spite of the strong interaction between the nucleons. I tried to explain this in a qualitative way as an effect of the Pauli principle, which prevents a nucleon from being scattered by its neighbors when all states into which it could be scattered are occupied by others. In a collaboration with F. Friedman, we tried to show how effects of strong interaction (such as a complete amalgamation of an incident nucleon with the target nucleus) and effects of quasi-free particle motion (such as the shell model results and the so-called direct reactions) can simultaneously occur without logical contradiction.

After many years of work in nuclear structure, the field became a little too complicated for me. I don't like complicated things, and therefore, I thought that high-energy physics would be the right thing to go into. In some ways I always was involved in high-energy physics. I mean, the difference between nuclear structure and high-energy physics is only 15 or 20 years old and nuclear physics was previously high-energy physics. The real reason for my changing fields was that after having written a big book I became an "expert." Pauli once said: "Don't become an expert, because of two reasons: First, you become a virtuoso of formalism and forget about real nature, and second, if you become an expert, you risk that you are not working for anything really interesting anymore."

After the Second World War, I became more and more involved in helping to establish contact and collaboration between scientists of different countries. I was convinced that an international scientific community existed. Scientists are able to understand each other quickly; their common interest and aim of obtaining deeper insights into nature establishes a firm bond that easily bridges geographical frontiers or political and social differences. The scientific frontiers, the scientific problems, are truly international—or "a-national"—and to a large extent are independent of the political and economic systems. Science can build bridges between the different parts of our divided planet for this reason. The years after the war were difficult in this respect; contacts were broken and many ideological and bureaucratic obstacles hindered a free exchange of ideas and free collaboration. I tried very hard—not always successfully—to bring scientists from different countries together and to overcome bureaucratic restrictions that prevented exchanges between East and West.

The ravages of World War II left their mark upon European science. The community was dispersed and many of its facilities were out of date or damaged. Something had to be done in order to reestablish opportunities for the tremendous European intellectual power to participate in the development of modern science as it did so successfully before. I tried to help. I often went to Europe and taught and discussed physics with my friends. In 1950 I was the first *professeur etranger* at the Sorbonne. In 1960, after the tragic death of J. Bakker, the director of CERN,[1] I was asked to replace him. This was the time when the newly constructed accelerators were ready to be used for research. Here was my opportunity to do something important. I accepted and for five years remained as the director general of CERN. Those years were among the most wonderful of my life. When I look back, those five years seem like 20 years of my life, because there was so much to do, because there was so much to learn, and because there was such excitement and exhilaration in the challenge of creating an atmosphere of intense collaborative research and discussion among all participants, in dealing with problems of human relations between different nations, different physicists, different professions, engineers, mechanics, workmen, and even politicians. And it was a great adventure to be involved in the planning of future European developments such as the proton storage rings and the 300 Gev accelerator. Why did I quit after five years? I don't know why I did it nor

[1]CERN, the Center for European Nuclear Research, is a large international research laboratory, located in the neighborhood of Geneva, Switzerland, and run by 12 countries and devoted to research in high energy physics.

whether I regret it. However it was interesting to come back to America with all those new experiences from CERN.

This account of my life as a physicist must stop here. But I do not want to give the impression that physics was more interesting and exciting in the past. True enough, the early years of quantum mechanics were unique in the sense that so much was understood and explained. But the rate of exciting new discoveries and new ideas has not slowed down. Just look back over the last 15 years of particle physics alone: the antiproton in 1955, the violation of parity in 1956, the discovery of baryon and meson resonances around 1960, the hadron symmetries and the prediction of the omega particle in early 1960, the second neutrino in 1962, the violation of the matter-antimatter symmetry in 1964, the discovery of some pointlike structure within the nucleon in 1968. These 15 years provide a rate as exciting and challenging as ever. Obviously, our theoretical insight into these phenomena has lagged, but we expected too much. We are apt to forget how much experimental material had to be accumulated before quantum mechanics was discovered. When we learn about these things today, we never hear about the tedious and torturous ways in which those insights were reached; we only learn the most logical and direct approach. Physics has not become less exciting; the lives of younger physicists are as interesting and exhilarating as mine has been and is yet. I hope that the study of nature will give them as much help and strength as it gave me in finding a sense and purpose in this difficult and problematic period of human history.

Part 1
Fundamental Questions

Part 1 contains three articles dealing with the fundamental concepts of quantum mechanics. The first article, "Quality and Quantity in Quantum Physics," deals with the wave-particle duality and with the modern explanation of atomic properties, in particular with the specificity and the stability of atoms. The concept of the quantum ladder is introduced that characterizes the recurrence of the same quantum principle in subsequent levels of the structure of matter; in molecules, atoms, nuclei, and in subnuclear phenomena. The second article, "The Quantum Ladder," an interview, serves as a further illustration of this concept. The third article, "Niels Bohr, the Quantum and the World," is devoted to the work and life of Niels Bohr, the founder of atomic physics. It tries to give a simple account in everyday language of Bohr's concept of complementarity, which not only plays a pivotal role in the conceptual fabric of modern physics but also acquires deeper philosophical significance when applied to other human situations.

Quality and Quantity in Quantum Physics

I

In this essay the antithesis Quality versus Quantity is understood to be related to the contrast between the specific and the unspecific, between individuality and continuous change, or between well-defined patterns and unordered flow. In this sense Quality and Quantity play a fundamental role in the basic concepts of quantum physics and in our scientific picture of the natural world. Of necessity our presentation of the role of this antithesis will be somewhat short and sketchy, and it can be justified only by the fact that not enough attempts are made by physicists to elucidate the basic ideas of quantum theory, a field of human thought that, more than any other scientific achievement, has deepened and broadened our understanding of the world in which we live.

A case in point is the history of ideas on the structure of the planetary system. Let us consider three phases of this history: the ancient Pythagorean ideas; the modern ideas based on Newton's theory of gravity; and our present ideas regarding another planetary system—the system of electrons revolving round the atomic nucleus, the atom.

We are not interested here in the details of the Pythagorean system of heavenly bodies (e.g., the question as to which is the center and which moves in circles around the center) but in only one feature: the fundamental importance the Pythagoreans attributed to the numerical ratios of the radii of the orbits and to the numerical ratios of the periods of revolution of the different heavenly bodies. They considered the simple numerical relations between these data as the essence of their system. According to their ideas these relations were the embodiment of the "harmony of the spheres"; they represented the inherent symmetry of the heavenly world as contrasted to the earthly. The harmonious interplay of the various celestial motions produced a music

Revision by the author of article originally in "Quantity and Quality," *Daedalus* **88,** 592, Fall 1959.

whose chords were audible to the intellectual ear and were a manifestation of the divine order of the universe. Thus not only the general structure of the solar system but also the specific shapes and the actual periods of the orbits themselves were significant and uniquely predetermined. Any deviation would have disturbed the harmony of the spheres and therefore was unthinkable.

This picture of the solar system did not survive the development of a better understanding of the underlying facts. Isaac Newton recognized that the phenomenon of gravitational attraction was the guiding principle by which the motions of the planets would be fully understood. This discovery was the end product of a development that led to a complete change in the attitude toward the problem of planetary motion. Not only did it become obvious that the sun was the center of the system, but it was also recognized that the motions of the planets were governed by the same laws as governed terrestrial phenomena. For our purposes here, the following is important: The laws of gravity admit of many ways a planet might circle around the sun; it can be any orbit of elliptical shape. The specific orbits in which our planets actually are found cannot be determined by the fundamental law of motion, but by so-called "initial conditions," those prevailing when the system was being formed. In this sense the actual shapes are accidental. Slightly varying conditions at the beginning would have produced different orbits. We now have good reason to believe that there are many other solar systems among the stars whose planets have orbits quite different from those in our own system.

Here we have a characteristic trait of physical thinking up to the advent of quantum physics, a period generally referred to as that of "classical" physics. The fundamental laws determine only the general character of the phenomenon; they admit of a continuous variety of realizations. The phenomena actually realized depend upon influences acting before the phenomenon was allowed to develop without further interference from outside. For example, if another star swept close by our solar system, the planetary orbits would undergo a thorough change and would be quite different after the star had left; similarly, if another planetary system with a star like our sun and planets of the same masses as ours existed, it would be highly unlikely that its orbits would bear any resemblance to the orbits of our own system, except that the orbits also would be ellipses and that their time of revolution would be the same function of the size of the ellipse.

It is typical of classical physics up to the turn of the century that its laws predetermine only the general character of the phenomena. The exact course of events can be predicted from the laws only if the situation at some past time is exactly known. The laws tell us only how an event develops; they do not tell us why we have this "solution of the equation" rather than another which would equally well fit the laws. This choice is considered accidental.

Since the time of Newton, classical physics has developed with the usual scientific crescendo and with ever increasing success, not only in mechanics but also by encompassing many different phenomena such as electricity and heat. The laws of nature discovered in this development were extremely successful in describing the character of many diverse phenomena. Hence these laws had to be a part of the real structure of the world around us. However, around the turn of the century it became obvious that the world of classical physics lacked some essential features to be found in the actual world. The stage was set for new discoveries.

II

To illustrate the situation at the beginning of quantum physics, let us return to a planetary system analogous to that of the sun and its planets—that of the atom. The properties of the solar system were exhaustively understood through the application of the laws of classical mechanics. The experiments of Rutherford and his contemporaries had shown that there exists another similar system, the system of electrons in the atom. It consists of electrons circling round an atomic nucleus, just as the planets revolve round the sun. The attractive force replacing gravity is the electric attraction between the negatively charged electrons and the positively charged nucleus. This force should produce the same type of motion in both cases, since it displays one important characteristic: It decreases as the reciprocal square of the distance.

The predictions on the basis of the atomic model were fulfilled in many respects. For example, the time of revolution of the electrons (which can be deduced from the frequency of the light emitted by the atoms) is just about what one would expect from the size of the orbits (as deduced from the atomic dimensions). However, the atom has some very important properties one would never expect in a planetary system—the most striking, the *identity* of the atoms of a given material. One must be impressed by the fact that pure materials show identical properties, no matter where they come from or what their previous history has been. Two pieces of gold, mined at two different locations and treated in very different ways, cannot be distinguished from one another. All the properties of each individual gold atom are fixed and completely independent of its previous history.

The identity of individual atoms is strikingly not in keeping with what is expected from a mechanical system, particularly one like the planetary system. The particular shape and size of the orbits are expected to depend markedly upon the past history of the system; it would be extremely improbable to find two atoms with exactly the same size and shape. The difficulty becomes obvious when we consider a gas such as air: The atoms in air collide many million times per second. According to classical mechanics, each of these collisions would thoroughly change the orbits of the electrons. In fact, however, the atoms emerge completely restored to their original form after each collision.

The problem of definite shapes in atomic phenomena versus the arbitrarily changing forms in classical mechanics permeates atomic physics. We find definite "qualities" in the atomic world where we expected quantitative differences. The crystal structure of matter reveals well-defined identical geometrical patterns in the atomic structure that ought to be absent in classical mechanics. Nature exhibits all around us characteristic and specific properties of various materials, which, in spite of their overwhelming variety, are always reproducible and recurrent. The specificity of material qualities in nature is in need of a fundamental explanation.

Even the existence of elementary particles such as electrons, protons, neutrons needs some better understanding. These particles are the building stones of the atoms and must *a fortiori* exhibit complete identity among members of one kind, if the atoms of a given type prove to be identical. Within the framework of classical physics, it is hard to understand why there should not exist electrons with slightly less charge, or with a different mass, or with a spin (rotation about an axis) somewhat at variance with the spin of the observed electron. It is the existence of well-defined specific qualities, in which nature abounds, that runs counter to the spirit of classical physics.

In this connection one must mention the Boltzmann paradox, although the fundamental significance of this point might escape the nonphysicist. There seems to be no end to the following regression: Matter consists of atoms; atoms of electrons and nuclei; nuclei of protons and neutrons; electrons, protons, and neutrons of. . . . As such, the existence of this regression should not worry us; it serves as a constant challenge to further research. However, in 1890 Boltzmann pointed out that on the basis of classical mechanics one is led to expect that for a system of atoms in thermal equilibrium at a given temperature, the thermal energy should be divided among all the modes of motion. This leads to a puzzle: All possible motions should share in the heat motion; if a piece of material is heated, the electrons should run around faster; the protons should vibrate more strongly within the nuclei; the parts of which the protons are made should move faster within their bounds; etc. Hence the above-mentioned regression would unavoidably lead to an infinite *sink* of heat energy, and it would need immense energies to heat the smallest part of matter. Here, as before, the classically admissible modes of motion are obviously too unspecific and too varied and do not explain the structure of matter.

One main feature of classical physics is the divisibility of each process. Every physical process can be thought of as consisting of a succession of partial processes. Theoretically at least, each process can be followed step by step in time and space. The orbit of an electron round the nucleus may be thought of as a succession of small displacements. The electron of a given charge may be thought of as consisting of parts of a smaller charge. This is the point to be discarded if one wants to understand what we see in nature: quality, specificity, and individuality.

III

The great step forward that solved the paradoxes here outlined was achieved within only thirteen years, from the discovery of the quantum orbits of the atom by Bohr in 1913 to the final development of quantum mechanics by Bohr, Heisenberg, Schrödinger, and Dirac in 1926. The idea of the quantum of action, however, had already been conceived in 1900 by Max Planck.

The study of the properties of atoms has led to many indications of new phenomena outside the scope of classical physics. The most striking is particle and wave duality. In classical physics a beam of light and a beam of electrons are fundamentally different. The former is a bundle of electromagnetic waves propagating through space in a certain direction. No material is moving; only the state of the electromagnetic field in space is changing. In contrast, a beam of particles consists of actual matter in small units moving straight forward; it is as different as is the motion of waves on a lake from that of a school of fish swimming in the same direction. All the greater was the surprise of physicists when electron beams were found to exhibit wave properties, and light beams to exhibit particle behavior.

The particle nature of light was revealed by the discovery of a granular structure of light; the energy of the momentum of the beam is transferred to matter in finite amounts—the so-called light quanta. The size of the energy quantum is proportional to the frequency f; it has the value hf, where h is Planck's constant. The existence of the smallest package of energy hf turns out to be a general property of any vibrating process.

The wave nature of particle beams manifests itself in many ways. One is the well-known observation that particle beams show the same kind of "interference" as wave beams. A beam that penetrates a screen through two slits shows the characteristic intensity patterns, which are quite different from the simple sum of intensities expected of two separate beams emerging from the slits on the basis of the classical picture of particles. The pattern of intensity is in fact the same as if obtained from a wave passing through two slits.

Another perhaps somewhat indirect but most fundamental manifestation is found in the atom itself. In many respects the electron orbits show a striking similarity to vibrating waves restricted within the confines of the atom. For example, a wave confined to a finite volume—that is, a "standing wave"—can assume only a certain restricted number of shapes, in particular when its frequency is supposed to be low, as it must be for the states of lowest energy according to Planck's law. These shapes are well defined and have simple symmetrical structures, a fact known from other examples of standing waves, e.g., those in a violin string or in the air column of an organ pipe. They also have the property of "regeneration"; when a perturbing effect has changed the shape, they assume their original shape after the perturbation is over. Here we find some of those new essential features that were missing in the classical picture. We find typical well-defined shapes, the shapes of vibrations that are assumed by an electron wave confined by the attractive force of the atomic nucleus. These shapes are universal and depend only on the symmetry and strength of the confining electric field. They are the fundamental patterns that matter is made of. This is why a copper atom is exactly the same wherever it is found and whatever its previous history.

These typical, ever recurring qualities of matter are based upon the wave nature of the electron. The questions must arise: How can a particle in motion exhibit any wave nature? How is it possible that an electron is partly a particle and partly a wave? After all, a careful tracing of the electron along its motion must decide this question and put it in either one or the other category. Here we come to the question of the divisibility of atomic phenomena. Can we really perform this tracing? There are technical problems in the way. If we want to "look" at the detailed structure of the orbit, we must use light waves with a very small wave length. Such light, however, has a high frequency, hence a big energy quantum. When it hits the electron it will knock it out of the orbit and destroy the very object of our examination. These considerations are the basis of Heisenberg's uncertainty relations. They express the negative statement that certain physical measurements are impossible. Characteristically, just those measurements which would decide between the wave or the particle nature of the electron (or proton, or any other entity) are impossible. If one performs these measurements, the subject has thoroughly changed its state by the very act of performance.

Here we recognize the highly important fact that this impossibility of certain measurements is more than a mere technical limitation that some day might be overcome by clever instrumentation. If it were possible to perform such measurements, the coexistence of wave and particle properties in a single object would collapse, since these measurements would prove one of the two alternatives to be wrong. However, we know from a great wealth of observations that our objects exhibit both wave and particle properties. Hence the Heisenberg restrictions must have a deeper root: They are a necessary corollary to the dual nature of atomic objects. If they were broken, our interpretation of the wide field of atomic phenomena would be nothing but a web of errors and its amazing success would be based upon accidental coincidences.

Atomic phenomena present us with a much richer reality than we are accustomed to face in classical macroscopic physics. The response of the object to our experimentation displays features that do not occur with objects in our macroscopic experience. Hence our description of the object cannot be as "detached" from the observing process as before. We can describe atomic reality only by telling truthfully what happens when we observe a phenomenon in different ways, although it seems incredible to the noninitiated that so many things should happen to one given object.

The wave nature of the electron in the atom is connected with the indivisibility, the wholeness, of the state of the atom. If we force a subdivision of the process and try to "see" more accurately where the electron "really" is within this wave, we find it there as a real particle, but we have destroyed the subtle individuality of the "quantum state." It is the wave nature, however, which gives rise to the characteristic properties of the quantum state—its simple shape, the regeneration of the original form after perturbation, in short, the specific qualities of the atom. The great discovery of quantum physics is the existence of these individual quantum states, each of which forms an indivisible whole, as long as they are not attacked by penetrating means of observation. Any attempt to observe subdivisions uses means of such high energy that they destroy the delicate structure of the quantum state.

The same situation exists also with the previously discussed electron beam which passes through a pair of slits in a screen and exhibits interference phenomena afterward. This phenomenon also has its individuality, its wholeness. When one tries to arrange a follow-up experiment in order to find out through which hole the electron went, the interference phenomenon is gone. The follow-up is too severe an operation, it destroys the wholeness of the quantum phenomenon.

At this stage of our discussion it will appear quite natural that predictions of atomic phenomena sometimes must remain probability predictions only. The prediction of the exact spot where the electron will be found after destroying a quantum state with high-energy light is a case of this kind. The quantum state is an individual entity which cannot be divided into parts without destroying it, although it spreads out over a finite region in space. If the quantum state is looked at with pin-pointing light, the electron will be found somewhere in the region of the wave, the exact point being undefined.

We now can return to the cause of identity among the same kind of atoms, and of their characteristic properties. The kind and shape of a standing electron wave is fixed and given when confined by the electric attraction of the nucleus, as is the shape of the vibration of a violin string. The standing wave of lowest frequency is spherically symmetrical, the next higher one has a "figure eight" symmetry; each step has its well-defined shape. These are the fundamental forms of which atomic structure is built. If we destroy an atom by removing an electron and later try to build it up again, the electron will return to the same quantum state from which it previously had been removed. There exists only one unique state of lowest energy for each kind of atom. This is in complete contradistinction to the situation in a classical planetary system.

We are reminded of the Pythagorean "pre-established harmony": The atomic quantum states have specific shapes and frequencies which are uniquely predetermined. Every hydrogen atom in the world strikes the same chord of frequencies, as given by the Balmer formula of spectral terms. Here we find the "harmony of the spheres" rediscovered in the atomic world, but this time clearly understood as a vibration phenomenon of confined electron waves. The complete identity of two gold atoms comes from the fact that the same number of electrons are confined by the same electric charge in the center and therefore produce the same wave vibrations.

It is often said that the atomic world is less "real" than the visible world around us because of the fact that we cannot describe the atomic phenomena independently of the mode of observation, and because of the fact that one uses dual descriptions which cannot be visualized in any simple way or calculated without the use of abstract mathematics. Heisenberg says:

> The conception of the objective reality of the elementary particles has thus evaporated in a curious way, not into the fog of some new, obscure, or not yet understood reality concept, but into the transparent clarity of a mathematics that represents no longer the behavior of the elementary particles but rather our knowledge of this behavior.[1]

[1] Werner Heisenberg, "The Representation of Nature in Contemporary Physics," *Daedalus* **87,** 100, 1958.

We do not agree with the claim that there is any lack of reality in the atomic world. After all, the visible real world consists of the same atoms that exhibit this strange behavior. It is true that the atomic world differs from our accustomed world more than anyone had expected; it has much richer patterns of phenomena than we can visualize with classical concepts. But all this does not make it less real. It is not very meaningful to distinguish between the actual behavior of the elementary particles and our knowledge of this behavior. It is precisely the ever growing insight into the detailed workings of nature which gives us confidence in having discovered something about the real world.

IV

The individuality and the stability of the quantum states have definite limitations. The atom has a unique and specific shape only as long as it is not disturbed by outside effects strong enough for an excitation of higher quantum states. Under very energetic interference from outside, the individuality of the quantum effects disappears completely and the system acquires the classical continuous character (often referred to as the correspondence principle). Hence the quantum character of mechanical systems is limited; it is exhibited only as long as the disturbing factors are weaker than the excitation energy to higher quantum states. This excitation threshold depends on the character of the system. It is always higher, the smaller the spatial dimension of the system. For example, it needs very little energy to change the quantum state of a large molecule; it needs much more to change the quantum state of an atom; and it needs many thousand times more energy to produce a change within the atomic nucleus. We arrive at a characteristic sequence of conditions which we may call the "quantum ladder."

At very low temperatures, the molecules of every substance form one big unit, a tightly bound crystal, in which one part is identical to any other. If we warm it to a higher temperature, melting or evaporation sets in and liquids or gases result. In a gas such as air at normal temperature, each molecule moves for itself in differing paths, bouncing against one another in irregular motion. The motions of the molecules are no longer alike; they are constantly changing, and they correspond to what we expect on the basis of classical mechanics. The molecules themselves, however, are still identical, one to the other. They interact as do inert billiard balls. The collision energies are not high enough to destroy their quantum state.

At still higher temperatures, the energy of collision surpasses the excitation energies of the molecules. The internal motion of the atoms and electrons participates in the exchange of energy. These are the temperatures at which the gas begins to glow and emit light. If still more energy is supplied, the molecules split into atoms, and further on the electrons are torn off the atoms. Then the atoms lose their individuality and specificity. Electrons and atomic nuclei move freely and in random fashion; no two electrons move exactly alike. This state of affairs occurs at temperatures as high as exist in the interior of stars. It is possible, however, to create similar conditions in the laboratory for a a small number of atoms. This is the object of "plasma physics." At those energies the atomic nuclei are still in their ground states. They are still identical and specific, whereas the atoms are already reduced from their specific qualities to unspecific random behavior.[2] Only if the energies of millions of electron volts are fed into the system, as is done in our big particle accelerators, are the higher quantum states of the nucleus excited or the nucleus even disintegrated into its constituents, the

[2]There is here a danger of confusion in our terminology: the word "quantum" is not related to what we understand by "quantity" versus quality. The term "quantum state" applies to the peculiar individual states of motion in atoms, molecules, or nuclei, which are the basis of the specificity and *quality* of these objects.

protons and the neutrons. Once this is done, the nucleus also has lost its quality and its specific properties, and has become a classical gas of protons and neutrons.

The newest giant accelerators are about to pour so much energy into the protons and neutrons themselves that the latter will begin to show internal structure and differentiation, and thus lose their innate identity. This development may advance toward new and unknown structures if the energy is further increased—or it may stop at some point, without yielding any new particles. We do not know and probably will never know unless we try it out.

The quantum ladder has made it possible to discover step by step the structure of the natural world. When we investigate phenomena at atomic energies, we need not worry about the internal structure of the nuclei; and when we study the mechanics of gases at normal temperatures, we need not worry about the internal structure of the atoms. In the former case we can consider the nuclei as identical unchangeable units—that is, as elementary particles; in the latter case each atom may be considered as such. Thus the observed phenomena are simpler and they can be understood without any knowledge of the internal structure of the constituents as long as the prevailing energies are so low that the constituents can be considered as inert units.

The phenomenon of the quantum ladder also solves the Boltzmann paradox. The finer structure of matter does not participate in the exchange of energy until the average energy has reached the level of its quantum excitation. Hence only those types of motion whose quantum energies can be excited at the prevailing temperatures participate in the heat exchange.

Let us now descend the quantum ladder, starting at the highest step known today. This may be a gas of protons, neutrons, and electrons at extremely high temperatures, with kinetic energies of many million electron volts. Not much individuality can be found under these conditions, except for the three elementary particles. Their motion is random and hence without any special order. At lower temperatures, say with kinetic energies of less than a million electron volts, the protons and neutrons assemble and form atomic nuclei. Much more specificity now enters into the picture. There are many possible atomic nuclei, the nuclei of the ninety-two elements and their isotopes, each a well-defined individual state. However, the motion of the electrons and the atomic nuclei is still at random, unordered, and continually changing. At still lower temperatures, corresponding to energies of a few volts only—this is the energy corresponding to the temperature of the surface of the sun (12,000 degrees F.)—electrons have fallen into regular quantum states around the atomic nuclei; this is the point on the quantum ladder at which the atoms with their specific individualities and chemical qualities appear. If we descend further, to the region of a tenth of an electron volt (about a few thousand degrees F.), we see that atoms can form simple molecules, and we find a much greater variety of chemical compounds, as distinct and specific as atoms, only somewhat less stable.

A further lowering of energy to a few hundredths of an electron volt (room temperature) brings us to a region where most molecules aggregate to liquids and crystals, thus adding to the diversity of matter. It is also the region in which giant chain molecules are formed. We have opened a completely new chapter of material specificity: living organisms. It starts with the formation of a great variety of chemical compounds of carbon with hydrogen, oxygen, and nitrogen—such as nucleic acids, amino acids, and proteins. The detailed dynamics of these giant molecules are not yet well understood, but some of their properties are well known. The most striking is the ability to include the formation of its replicas by combining simpler molecules into the pattern of the macromolecule itself.

The possibility of reproduction brings about a new mechanism: The structure most suited to reproduction, the one best protected against damage, will reproduce itself most abundantly. Hence we get a chain development of structures, the living organisms, which become consecutively better adapted according to the mechanism of natural selection. The reproduction of living structures is determined and guided by certain large molecules, of which the most important is DNA (desoxyribonucleic acid). The internal structure of DNA (in particular the order in which the purine and pyrimidine bases are arranged in it) is the determining factor for the properties of the units which are constantly reproduced in the cycle of life. Hence it is again the individuality of quantum states which is responsible for specificity in life. The specific structure of the nucleic-acid bases and the stability of the order in which they are arranged in DNA form another example, albeit a complicated one, of unique and identical quantum states. Because of the large length of the macromolecules, the number of possible quantum states is enormously greater than in the case of simple atoms or molecules, and their forms are much more intricate and complicated. This is reflected by the great variety of living species.

The existence of life requires that the temperature must be low enough to allow the formation of the macromolecules, but it also requires temperatures high enough for the supply of energy necessary for life processes. If we proceed downward on our quantum ladder to zero temperature, life decays, and all matter forms a big crystal in which many of the existing varieties are preserved but are frozen into inactivity. Everything is then found in its lowest state, a state of high specificity but without any change or motion. This is the stage of death.

Very probably the development of matter in the history of the universe has descended the quantum ladder just as we have described it, from high to low energies, adding new quality with each step. The history of the material world as we see it immediately around us probably began in some accumulation of protons, neutrons, and electrons of very high energy, compressed by forces of gravity, within a young star. This was a period of little differentiation. Later on, the elementary particles aggregated to atom nuclei, and in the colder regions of the star atoms were formed. This was the first step toward quality and organization. Individual properties began to appear, motion and radiation were no longer all uniform. Classes of identical objects were created—one thing could be distinguished from another.

On the surface of stars and the colder planets, the temperature further decreased, and conditions suitable for the formation of a great variety of chemical compounds were established. At that stage the world acquired an aspect not unknown to us, one of rocks, deserts, and waters, abundant in minerals and chemicals but without any form of life. Finally, at certain places in the universe where conditions were favorable, the great adventure of nature took place of which we ourselves are a part. The organic macromolecules began their cycles of reproduction, and evolution toward the varied forms of life appeared. The development from quantity to quality then reached that stage of diversity and abundance which we know as the world in which we live. Human life, men's thought, and men's feelings are but one manifestation of this stage.

The contrast to the formless chaos of the beginning vividly illustrates the innate trend of matter toward distinction and specificity, a trend ultimately based upon the stability and individuality of quantum states. We who are living in the twentieth century are privileged to witness the most exciting phase of this development: the moment when nature in its human form begins to recognize a few of its own essential features.

The Quantum Ladder

Quantum physics is very different from classical physics. How do you see the difference?

I like to say it in the following way: Before we got to quantum theory our understanding of nature did not correspond at all to one of the most obvious characters of nature, namely the definite and specific properties of things. Steam is always steam, wherever you find it. Rock is always rock. Air is always air. This property of matter whereby it has characteristic properties seems to me one of the most obvious facts of nature. Yet classical physics has no way of accounting for it. In classical physics, the properties are all continuous.

What do you mean by "continuous"?

There are no two classical systems that are really identical. Take the planetary systems of stars, of which we all know that there are billions. According to our present knowledge, you can be sure that no two of them are exactly identical. In some, the sun will be a little larger, in some the planets would be a little larger, the orbits would be a little different . . . . Why? . . . Classical physics allows us an immense range of possibilities. The behavior of things depends on the initial conditions, which can have a continuum of values.

Now quantum theory changes this fundamentally, because things are quantized. No longer is "any" orbit possible, only certain ones, and all the orbits of a particular kind are the same. Thus in quantum theory it makes sense to say that two iron atoms are "exactly" alike because of the quantized orbits. So, an iron atom here and an iron atom in Soviet Russia are exactly alike. Quantum theory brought into physics this idea of identity.

I'm struck that you stressed the word "exactly," because for many people it's a certain inexactness that characterizes quantum physics. They remember the uncertainties.

Revision by the author of an interview, "Thinking Ahead with Victor Weisskopf," in *Int. Sci. Technol.*, **62,** June 1963.

I'm an old fighter against this interpretation of the uncertainty relation. Quantum theory brought in just that exactness.

The classical Greek approach was apparently based on sound intuition. The Greeks had a picture of discreteness in nature.

Yes, but the Greeks postulated the existence of atoms; they did not explain it. One cannot understand on the basis of classical physics how it is possible to have a mechanical system of one kind and a mechanical system of another kind and no mechanical system in between.

So that atoms, which were axiomatic to the Greeks, remained unexplained assumptions through the nineteenth century. You had experiments on atomic weights and had the kinetic theory of gases but did not have atoms understood?

They remained axiomatic up until 1913. If you look a little under the surface, you hit the same problem always. You mentioned gas theory. Now before quantum theory, people rightly looked at gases as colliding atoms. Yet how come such collisions don't change the nature of atoms? Atoms must have a structure, a mechanism, inside, and the collision must leave some change in it. Yet we know that it isn't so. The stability of the atom is something that is not understandable in classical theory.

Was this question posed at all before 1913?

Oh yes, in the famous Boltzmann paradox: Classical mechanics leads you to expect that, for a system of atoms in thermal equilibrium at a given temperature, the thermal energy should be shared equally among all the possible modes of motion. *All* modes. In a piece of heated material the electrons should run around faster, the protons should vibrate more rapidly within the nuclei, the parts of which the protons are made should vibrate more strongly within their bounds, etc. Thus the specific heat in any ordinary piece of matter should be extremely large. In actual fact the specific heat has just the size that can be accounted for by the external motions of the atoms alone. It was not understandable how the heat energy doesn't get into the atom and excite the internal degrees of freedom. This Boltzmann paradox came in 1890, well before quantum theory. There was no explanation.

Your implication is that there had already been advance beyond the Greek idea of the atom as the uncuttable one. Did the nineteenth-century physicists imagine there was structure within the atom?

Yes. The atom emits light, and after the discovery of the electromagnetic nature of light, it was clear there must have been some motion inside the atom that emits the light, so there must have been internal structure. And there was also a philosophical idea behind it; namely, the concept of an imaginary atom without internal structure doesn't make much sense. One must ask: "What's inside?" Now, one could have said it is solid, but even if it is solid, we know the solid has a structure.

So the philosophical question of what happens if you cut the atom remains.

To my mind, quantum theory for the first time indicates how one has to deal with problems of this kind. Quantum theory tells us that an atom is a nondivisible entity, *if* the energies applied to it are below a certain threshold. If the processes inflicted upon the atom are below a certain threshold, the atom is really indivisible, in the real sense of the word. It means that if atoms collide with energies less than the threshold, they bounce off completely unaffected, in exactly the same state that they were before. This is the new idea. That's the quantum idea.

However, when you are way above the threshold, the atoms go to pieces, and they behave like ordinary classical systems containing parts and particles. For example, at very high temperatures an atom is completely decomposed into its parts, the nucleus and the electrons. Consider a sodium atom and a neon atom. The former has 11 electrons, the latter 10. Below the threshold they are in their characteristic quantum states; very different. One is a metal, the other a gas. Above the threshold—at high temperatures—they are both a gas of nuclei and electrons. It is what one calls a plasma. In this state there is not much difference between a sodium plasma and a neon.

And just as there are threshold energy levels for disrupting an atom, so there are levels above which the nuclei would be split.

True enough, there is also a threshold above which the nucleus goes to pieces. This threshold is much higher than the atomic threshold. The atomic thresholds are of the order of a few electron volts; the nuclear threshold is at much higher energies—a few million electron volts.

I like to use the term "quantum ladder" for this. These are two steps of the ladder.

The quantum ladder has made it possible to discover the structure of the natural world step by step. When we investigate phenomena at energies characteristic of atoms we need not worry about the internal structure of their nuclei. And when we study the behavior of gases at normal temperatures and pressures we need not worry about details of the internal structure of the atoms that make up the gas. In that way the quantum ladder solves the Boltzmann paradox. The finer structure of matter does not participate in the exchange of energy until the average energy has reached that rung on the ladder.

Our whole experience in daily life is down low on your ladder, within the atomic level of the quantum ladder.

Yes. That or even lower. (See Fig. 1.) I started with the atom, but there are also steps farther down the quantum ladder, which are important for our life. Molecules, macromolecules, crystals. All life consists of macromolecules. The lower you go on the quantum ladder the more pronounced becomes the specificity of the structures: nucleus—atom—molecule—macromolecule—life.

How do these specific structures come about? This is a central point, it seems.

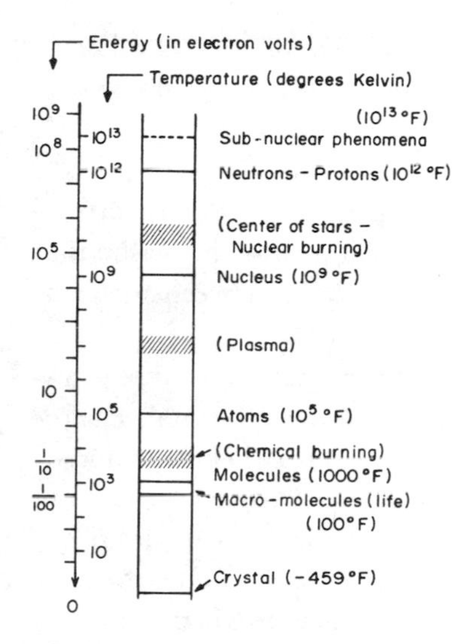

Figure 1.
Quantum ladder.

The quantum is an important precondition of the structuralization of nature. The particles fall into definite patterns, the quantum orbits of the nucleus, the quantum orbits of atom, the quantum orbits of the molecule, the quantum orbits of the macromolecule. Our hereditary properties are nothing else than the quantum states of the parts of a nucleic-acid chain, the so-called DNA. In some way the recurrence every spring of a flower of a certain shape is an indirect expression of the existence of certain quantum orbits in the DNA molecule—a consequence of the identity and uniqueness of quantum orbits.

Is it an accident or is there some deeper reason for the spacing of rungs on the quantum ladder?

Oh, yes, there is a very good reason. It lies in the size-energy relation: the smaller the system is, and the larger the mass of the moving particles is, the higher the quantum energies are. For example, it is not an accident that the quantum energy of the outer electron shell in an atom is only a few volts, whereas the quantum energy of a nuclear system is a few million volts. It's because of its smaller size and larger mass.

But, there must be a reason that atoms and nuclei have that size.

Well, the reason the electron shell exists is the electric attraction between the nucleus and the electrons, and the reason that the nucleus exists as a unit is the nuclear force between nucleons, that is, protons and neutrons.

Up to 1930 we dealt with two forces in nature—namely gravity and electricity. They are very well known to us on a macroscopic level. Bohr had found in 1913 that the chemical forces—the forces within and between the atoms—are electric. Only in 1930 when we first experimented on the inner structure of the nucleus did a new force come in, the nuclear force. It is the force that holds the protons and neutrons together when they form the atomic nucleus.

So the answer on the particular level of your question is, I think, the existence of these two force fields. Why these force fields? That is an unsolved question, and a question which I have a very definite feeling will be solved in our high-energy research.

High-energy research? Is there a step on the quantum ladder higher than the nucleus?

Yes, there is. Recently experiments with big accelerators have shown us that the proton and the neutron have a structure too. These particles can be changed into different states, they can absorb energy; in short, a world within the proton has been discovered. That is the next higher rung of the ladder.

What is the energy threshold for that step?

You can get at it only if you go to energies way beyond the temperatures and energies in the center of the stars. We don't know where the universe displays such energies . . . well, we have it in cosmic rays—very rare events. We have it at the target of our big accelerators, and maybe the center of the galaxy is of such type. We just don't know.

With the 30-billion volt machines you have just broken into a new highest level of energy. The immediate effect has been the discovery of 30 or 40 particles. The new machines seem to cause more confusion than anything else.

I don't accept your premise. The statement that high-energy physics has found 30 or 40 particles has brought this field into disrepute. But that reputation is wrong—for several reasons. One is that everybody counts the anti-particles as extra particles, which is as if you would double the number of animal species by calling the mirror image of an animal another animal. And there's more than that. I think it is wrong to call an excited state of a system, a new system. It's as if we would say that the excited hydrogen atom is another atom. It's becoming clearer and clearer that many of these particles are nothing else than the excited states of other particles.

Which particles are fundamental and which are excited states?

For example, the sigma particle, the lambda particle, and the xi particle are all excited states of the proton. I would go so far as to say we have only two elementary particles, the baryon and the lepton, and these particles have different states, different configurations, just like the hydrogen atoms. The proton and neutron are two states of the baryon, just like the spin up and spin down of the electron in the ground state of hydrogen. The lambda, the xi, and the sigma are excited states. The systematics of all these new states is what I like to call the third spectroscopy. We have atomic spectroscopy, the quantum levels of the atom, nuclear spectroscopy, the quantum levels of the nucleus, and now we have the third spectroscopy, which is the quantum levels of the nucleon.

The leptons also occur in different forms: as electrons, as neutrinos, and as heavy electrons (sometimes called mu mesons).

Where do the pi mesons and the K mesons fit in your picture? Aren't they also particles?

I would rather not call them particles. They are field quanta.

Just as the light quantum is a quantum of the electromagnetic field, so the pi meson and the K meson and the many other mesons are quanta of the nuclear field.

What is a field?

Fields began as a way of expressing the force between particles. The attraction between two unlike electric charges can also be expressed in terms of the action of the field of one on the other. But the field is not just a mathematical fiction, it is as real as its particle sources, and we can speak of its energy, etc.

Now, every field has a quantum. When a field propagates in space, when it is emitted in the form of radiation, it propagates in the form of quanta. The very best-known field quantum is the photon, the quantum of the electromagnetic field. The nuclear force seems to require several quanta—the pi meson and the other mesons. They all play important roles. The pi meson is responsible for the outer reaches of the nuclear field, and the other mesons are responsible for the force at closer distances.

You're not troubled that there are many quanta for this field?

I'm troubled, but not as much as one might think. The field is just somewhat more complicated. The Coulomb field falls off inversely with distance, it's just straight $1/r$; you see, the nuclear field is a complicated field, so no wonder one needs many quanta. What is complicated is that the quanta carry isotopic spin and another quantum number which is called hypercharge or "strangeness."

It's not just like you had quanta of that field, but also you have to adduce new quantum numbers.

Yes. These quantum numbers play an important role in the "third spectroscopy" we mentioned before. The excited states of the baryon can be classified according to these new quantum numbers.

But the basic idea is the same as in the spectrum of excited states of any atom, say hydrogen. The ground state is of course the proton or the neutron—the neutron has a little more energy. Then there are several excited states—the names originally given them make no sense any more so we call them by their various quantum numbers.

There's a nice historical parallel, isn't there? Just as the early history of optical spectroscopy was marked by the naming of spectral lines according to their appearance, so the names of nuclear particles have grown with their experimental discovery.

Exactly . . . now there are also excited states which have a different strangeness quantum number. And they have names—the lambda, for example. But they have names only because of the fact that their different strangeness makes them metastable and they last long enough to be observed as an apparently different particle. It's like an atomic state which is metastable.

You get these extreme energies from CERN and Brookhaven accelerators at intensities that apparently do not exist in the universe otherwise. When you are doing physics in an area that nature doesn't reach to, what are you doing?

I am deeply convinced that nature has such a variety that any process we find on earth will be of importance somewhere. And that's why I think that the experiments we are now doing on these highest rungs of the quantum ladder will have significance in one of those unsolved problems, such as the problem of the expansion of the universe, the creation of matter, or the fundamental structure of matter. It may be that the problems of the creation of the universe are connected.

You are now perhaps experimenting with conditions which, 20 billion years ago . . .

That is a matter of interpretation. As you know there are two views—which I would like to call two "religions." One is the Big Bang theory where the universe started billions of years ago with tremendous pressure and energies in a small volume, and the other is the Continuous Creation theory. I'm sure that the true answers to the questions would be neither one nor the other. But it is correct to think that if the Big Bang has something to do with reality, some of the early phases might have had to do with the latest rung on the quantum ladder.

You say "religion" to indicate a real difference between those parts of physics which have the authority of physical experiment and those parts that are extrapolating to conditions that in no sense can we duplicate.

Exactly. One must really draw a serious line there. Although I also believe that these speculations are the most exciting one can imagine. But they are really different from physics itself. Sometime, when we shall know more about these things, they might become true physics.

Is the line of progress inevitably toward the still higher and higher levels of energy; or does your ladder have a top rung?

A last quantum rung? I cannot tell. It is always the highest hope, maybe a hope only, of physicists that at the next step of the quantum ladder you will find the all-embracing principle. Heisenberg thought so. In every physicist there is an element of belief that you will sometime come to the recognition of some fundamental facts which close it, from which you can explain everything. I'm not so sure. It may be that nature is inexhaustible. But it might not be. How do we know?

You're saying there would be a last step if you could know what would happen if you used 100 times as much energy?

Yes. But in order to reach that state of affairs we must also have a Heisenberg or a Bohr of the future who gives us the theory that explains all phenomena in terms of what we know. Until we have that theory, we will never have a guarantee that there is not a new world coming up. We would have to build higher energy accelerators to find out.

Earlier you said that atoms are philosophically unsatisfying because you can always ask what is within them. If physics is not inexhaustible, then at some point you will truly have elementary, nondivisible particles.

I wouldn't call it the elementary particle, it might be something else. A field, or even some new thing which is as far from the field as the field is from the particle, consequently something new, but that embraces the whole. What it will be we don't know—we are just at the beginning.

Niels Bohr, the Quantum and the World

One achievement that stands out in the complex history of ideas of the twentieth century is the development of our concepts of the structure of matter. It was a steady development that penetrated deeper and deeper into the inner structure of the atom, ever widening our understanding of material things. Modern scientific progress is often described in terms of revolutions and upheavals, where one new theory destroys the previous theories. However, this description overlooks the fact that scientific development is intrinsically evolutionary. Any one of the new and so-called revolutionary ideas in modern science was a refinement of the old system of thought, a generalization or an extension. Relativity did not do away with Newton's mechanics—orbits of satellites are still calculated with Newton's theory—it extends its application to extreme velocities and establishes the general validity of the same concepts in mechanics and electricity. Quantum theory was perhaps the nearest to a revolution, but even these ideas, such as the uncertainty principle, must be considered as a refinement of classical mechanics for the application to very small systems; they do not change the validity of classical mechanics for the motion of bodies of larger size.

The steady and incessant growth of our understanding of material structure may have helped to steady the minds of the scientists who live in this century of turmoil and upheaval. It did not have that influence on society as such. Any growth of this kind of knowledge necessarily brings along more and more ways and means of dealing with new materials, new forms of energy, and new forces that can be turned on and off at will. This again, necessarily, changes the quality of life at an ever increasing rate, leaving us at odds with our accepted value system when we face the human problems created by the new developments.

Nothing could be better suited for an illustration and exempli-

Revision by the author of article originally in *N.Y. Rev. Books* **8,** 7, April 20, 1967.

fication of these problems than a study of the life of Niels Bohr. Here was a great physicist, one of the greatest, his name ranks beside Galileo, Newton, Maxwell, and Einstein. His contributions started the great development we are talking about and kept it going for half a century. To a greater degree than any other scientist, he was involved in the human problems of his science, in the impact on society and politics. He was born in 1885; his life as a scientist began about 1905 and lasted for fifty-seven years. 1905 was the year when Einstein published his first paper on special relativity; it was only a few years after Planck's discovery of the quantum of action. Bohr had the great luck to be present at the beginning, or perhaps mankind had the great luck of having him at that turning point. What a time to be a physicist! He began when the structure of the atom was still unknown; he ended when atomic physics reached maturity, when the atomic nucleus was put to industrial use for the production of electric power, to medical use in cancer treatment, and, unfortunately also, to military and political use.

The work of Niels Bohr can be divided into four periods. In each one he exerted a tremendous impact on the development of atomic science, in four different ways, at four different times. The first is the decade 1912–1922, from his meeting with Rutherford until the foundation of his famous Institute of Theoretical Physics in Copenhagen. In this period Bohr introduced the concept of *quantum state* and created an intuitive method of dealing with atomic phenomena. In the second period, from 1922 to 1930, he gathered around him in his new Institute a few of the most productive physicists of the world who, under his leadership, developed the ideas of quantum mechanics. This is the conceptual edifice that replaces his original intuitive method and gives an adequate description of the inner workings of the atom. The third period, 1930–1940, was devoted to the application of the new quantum concepts to electromagnetic and later nuclear fields and the exploration of the structure of the atomic nucleus. Then came the Second World War and the last period of his life, in which he acted as the great leader of physics, who was deeply concerned and involved in the social, political, and human consequences of the new discoveries.

Let us return to the first period, which began with the publication, in the year 1913, of his work on the quantum orbits of the hydrogen atom. This remarkable paper proposed to explain the unexplained properties of the atom by introducing a completely new concept into physics, the concept of quantum state. His ideas were based upon previous work by Planck and Einstein. He applied the idea of the quantum to the structure of the atom. There is hardly any other paper in the literature of physics from which grew so many new theories and discoveries.

This famous paper marked the beginning of a series of new insights. In the ten years following its publication, many things not previously understood fell into place: the structure of the spectra of elements, the process of absorption and emission of light, the reasons for the periodic system of elements, the puzzling sequence of properties of the 92 different atomic species. It was the period in which quality, the specificity of chemical substances, was reduced to quantity, to the number of electrons per atom. All this rested on Bohr's assumption of the existence of quantized orbits in the atom, at that time still a provisional hypothesis. Bohr's contemporaries, however, took it quite literally, although Bohr warned them in his papers and at meetings that this could not be the final explanation, that there was something fundamental to be discovered in order to understand really what was going on in the quantization of the atom.

The second period was the time in which the quantum was fully understood. It was a heroic period without any parallel in the history of science, the most fruitful and most interesting one of modern physics. There is no single paper by Niels Bohr himself that characterizes this period as did the 1913 paper in the first period. Bohr found a new way of working. He did not work as an individual alone; he worked in collaboration with others. It was his great strength to assemble around him the most active, the most gifted, the most perceptive physicists of the world. At that time, we find with Bohr at his famous Institute for Theoretical Physics, in Copenhagen, people such as Klein, Kramers, Pauli, Heisenberg, Ehrenfest, Gamov, Bloch, Casimir, Landau, and many others. It was at that time, and with those people, that the foundations of the quantum concept were created, that the uncertainty relation was first conceived and discussed, that the particle-wave antinomy was for the first time understood. In lively discussions, in groups of two or more, the deepest problems of the structure of matter were brought to light. One can imagine what atmosphere, what life, what intellectual activity reigned in Copenhagen at that time. Here was Bohr's influence at its best. Here it was that he created his style, the *Kopenhagener Geist,* the style of a very special character that he imposed onto physics. We see him, the greatest among his colleagues, acting, talking, living as an equal in a group of young, optimistic, jocular, enthusiastic people, approaching the deepest riddles of nature with a spirit of attack, a spirit of freedom from conventional bonds, and a spirit of joy that can hardly be described. As a very young man, when I had the privilege of arriving there, I remember that I was taken a little aback by some of the jokes that crept into the discussions, and this seemed to me to indicate a lack of respect. I communicated my feelings to Niels Bohr and he gave me the following answer: "There are things that are so serious that you can only joke about them."

In this great period of physics, Bohr and his men touched the nerve of the universe. The intellectual eye of man was opened to the inner workings of Nature that were a secret up to this point. Once the fundamental tenets of atomic mechanics were settled, it was possible to understand and to calculate almost every phenomenon in the world of atoms, such as atomic radiation, the chemical bond, the structure of crystals, the metallic state, and many others. Before that time, our environment was full of different forces: electromagnetic, cohesive, capillary, chemical, and elastic; then, all these forces were reduced to one —to the electromagnetic force. In the course of only a few years, the basis was laid for a science of atomic phenomena that grew into the vast body of knowledge known to us today.

A few words are in place to sketch the impact of the new ideas spurred by Bohr. Before these years, chemistry and physics were wide apart. Chemistry, on the one hand, was the science of matter and its specific properties. The atom was a concept of chemistry—the atom of gold, of oxygen, of silver: Different specific entities whose existence was noted but not understood. Physics, on the other hand, was a science of general properties of motion, of strain and stress, of electric and magnetic fields, and the two sciences were far apart. One was not yet able to answer the question: "Where do the specific properties of matter come from?"

The specificity of the atoms was a great miracle. What prevented Nature from producing a gold atom that is slightly different from another? Shouldn't there be intermediate atoms that are not quite gold but halfway to silver? Why couldn't there be a continuous change from gold to silver? What keeps all atoms of one species so exactly alike; why are they not altered by the rather rough treatment they suffer when the material is heated or subject to other outside influences? This question was even more acute and disconcerting when Rutherford found out that the atoms are little solar systems with the atomic nucleus as the center sun and electrons circling around it as planets do. Such systems should be extremely sensitive to collisions and other perturbing influences.

Bohr saw that there was a connection between these atomic properties and quantum theory. He tried to formulate the situation by postulating the existence of quantum states for atomic systems that are characteristic for each species. Electrons can assemble around the nucleus only in a few well-defined modes—the quantum states—and not in others. The mode that has the lowest energy is the one in which the electrons would invariably assemble under normal conditions. It is a stable configuration because any change would be possible only if enough energy can be supplied to reach the next quantum state, which is a definite step higher in the energy scale. It is that configuration that is responsible for the typical properties of the atoms.

So far this is nothing but a suitable formulation of the strange fact that atoms have specific qualities. It was somewhat more than that, because Bohr gave some rules for how to calculate correctly the energy of these quantum states in some simple cases. But the real significance of this new concept emerged when it became clear that it is intimately connected with the dual nature of electrons, whose motions are sometimes observed as a particle in motion, sometimes as a wave motion. The quantum states of atoms turned out to be nothing else but the specific vibrations of electron waves confined by electric attraction to a space close to the nucleus. A most exciting situation: the specific atomic states as harmonic vibrations of electron waves under the confining influence of the electric force of the nucleus. The specific properties of the elements are based upon a natural interplay of vibrations. The old dream of "harmony of the spheres" seemed to be revived.

But this situation was also deeply disturbing. How can it be that electrons exhibit wave and particle properties at the same time? There seems to be an irreconcilable contradiction between electronic particles revolving around the nucleus and vibrating electronic waves. Obviously, the mere existence of atoms with well-defined specific properties is already a proof that strange things must go on in the atom in order to make little solar systems of electrons exhibit such behavior. The discovery of the wave-particle nature of the electron only reinforces this. The main point in the Bohr approach to this problem lies in the refutation of the following view: Why not solve the whole problem by looking into the structure of the atom? After all, we could make use of the finest means of observation to find out what the detailed structure is like, and this ought to decide whether the electron is a wave or a particle.

Nature is arranged in a way that makes this approach impossible, because no observation of a tiny object can be made without influencing it. The quantum state has a peculiar way of escaping ordinary observation, because the very act of such observation would obliterate the conditions of its existence. The quantum state is a form of motion that cannot be divided into parts and followed up point by point, as we do when we describe the motion of a planet around the sun. It must be considered in its characteristic entirety and indivisibility. The quantum properties can unfold only when the atom is left undisturbed, when the perturbations to which it is exposed contain less energy than would be necessary to cross the threshold into the next quantum state. Then we find the atom with its characteristic properties, and it behaves like an indivisible entity. When we try to look into the details of the quantum state by some sharp instrument of observation, we necessarily pour much energy into it, and we destroy the quantum state. In fact, when an atom is given a large amount of energy, then it behaves like an ordinary solar system. The characteristic quantum properties are lost. The necessary coarseness of our means of observation—light comes in quanta and so does any other form of energy—makes "exact" observation in the old sense impossible. This is the basis of the famous uncertainty principle formulated by Heisenberg in 1926, when he was working with Bohr.

The quantum state represents a novel state of matter that cannot be described in the old-fashioned way. It exhibits features that do not occur with objects of our ordinary experience. This is why we must use more abstract terms when we describe atomic reality. It may seem incredible to the non-initiated that an electron behaves in certain situations like a wave and, in others, like a particle. But this is just part of the reality we are facing in the world of atoms.

Bohr introduced the term *complementarity* for this complex state of affairs. He was so fascinated with this new mode of arguing that he tried to apply it to some other aspects of human thought. For example, the problem of free will can be looked at in a similar way. The awareness of personal freedom in decision is an experience as clear and as factual as any other. When it is analyzed, however, by following up each step of decision making in its causal connection, the phenomenon of free decision is no longer apparent. A related complementarity is found in the well-known paradox of thinking about the thinking process and also in the juxtaposition of reasoning and acting. We can never analyze the actual process of the thinking that does the analysis; we can never act if we constantly think about the possible consequences of our acts; we have no time to reason during the process of acting. The legal and the humanitarian approach to a human conflict often shows similar features of contradiction that should be resolved by a complementary approach.

In a more jocular vein Bohr, being an enthusiastic skier, sometimes used the following simile, which will be understandable only to fellow skiers: "When you try to analyze a christiania turn into all its detailed movements, it will evanesce and become an ordinary stem turn, just as the quantum state turns into classical motion when analyzed by sharp observation."

One of the most debated applications of complementarity was

Bohr's attempt to formulate the problem of life in this light. In a celebrated talk given in 1933, he ventured the idea that the phenomena of life on the one side and the validity of physics and chemistry on the other are contradictions that could be seen in the light of complementarity, in the sense that any attempt to verify in all details the validity of physics in a living cell would necessarily kill the cell and destroy the very object of investigation. Thus, a new and different state of matter might exist, which never would lead to a situation at variance with the laws of physics, but still would be outside of their valid application. The recent developments of biology have made Bohr's idea somewhat less attractive. It has turned out that the phenomena of life may not be in such an irreconcilable contradiction to the laws of physics as the quantum state was to classical physics. Bohr gave a second talk shortly before he died, in which he revised his original ideas to some extent.

We have spent much time in commenting upon the first and second periods of Bohr's life. In fact, he reverted to the idea of complementarity all through his life, trying to formulate it in new and better ways. In the third period he turned to nuclear physics. This was also a most interesting development in modern physics. The atomic properties were explained by the peculiar vibrations that the electron waves perform when they are confined close to the atomic nucleus by the electric attraction between the nucleus and the electrons. The same phenomenon was found to reappear again within the nucleus, albeit on a much smaller scale of size and on a much higher scale of energy. The nucleus is a system of protons and neutrons held together by a strong nuclear force. Again we find quantum states of the nucleus with characteristic properties based upon vibrations of waves: This time it is a matter of neutron and proton waves confined by the nuclear force. Nuclear structure presents an impressive evidence for the general validity of quantum mechanics. Here the physicists face a repeat performance of the same principles on a new level, with a few characteristic differences: In the atom the nucleus dominates the electron motions or vibrations because of its big charge and its heavy weight. In the nucleus we face a republican regime, because all constituents have equal mass and exert an equal force on each other. Bohr was the first to analyze the typical properties of nuclear quantum states that stem from this difference. It would be too awkward to discuss this topic here; nevertheless, from the point of view of the historian of science, the following point should be mentioned: The weight of Bohr's personality was so great that for a decade those typical differences were in the center of interest, whereas the similarities between nuclear and atomic quantum states were left aside; only as late as 1950 Mayer and Jensen discovered the shell structure of nuclei in analogy to the shell structure of atoms.

The fission of uranium was discovered when Bohr was deeply involved in his studies of nuclear structure. Obviously this phenomenon captured Bohr's interest, and he wrote a fundamental paper on this process with the American physicist John Wheeler, a paper that had a decisive influence on the development of nuclear energy.

When Bohr set out for a new problem, he always found a "victim" among the younger physicists who happened to be in Copenhagen. This lucky man had the privilege to work with him day and night; Bohr tried to explain his ideas to the victim until they became clear to both of them. The most important part of such collaboration was the writing of a common paper on the subject. The victim was supposed to write down the sentences while Bohr dictated. Days went by, each sentence was worked at until it expressed the desired idea. The ideas took their real shape only during those attempts of formulation. The interaction between thought and language always fascinated Bohr. He often spoke of the fact that any attempt to express a thought, involves some change, some irrevocable interference with the essential idea, and this interference is all the stronger, the clearer one tries to express oneself. Here again there is a complementarity, as he frequently pointed out, between clarity and truth, between *Klarheit und Wahrheit* as he liked to say.

This is why Bohr was not a very clear lecturer. He was intensely interested in what he had to say, but he was too much aware of the intricate web of ideas, of all possible cross connections; this awareness made his talks fascinating but hard to follow. There is a delightful story that Bohr liked to tell in this connection, about a young man who was sent by his own village to another town to hear a great rabbi. When he returned he reported: "The rabbi spoke three times; the first talk was brilliant: clear and simple. I understood every word. The second was even better: deep and subtle. I didn't understand much, but the rabbi understood all of it. The third talk was a great and unforgettable experience. I understood nothing and the rabbi himself didn't understand much either."

The work on uranium fission inevitably brought him into a realm where physics and human affairs are hopelessly intertwined. But even before these discoveries, he was deeply aware of human problems. He was unusually sensitive for the world in which he lived. He was aware earlier than many others that atomic physics would play a decisive part in civilization and in the fate of mankind—that science cannot be separated from the rest of the world. The events of world history brought home this point earlier than expected. By the 1930s, the ivory tower of pure science had already been broken. It was the time of the Nazi regime in Germany, and a stream of refugee scientists came to

Copenhagen and found help and support from Bohr. He asked some to stay with him at that time; James Franck, Hevesy, Placzek, Frisch, the writer of this essay, and many others found a haven in Copenhagen where they could pursue their scientific work. But not only this, Bohr's Institute was the center for everybody in science who needed help, and many a scientist got a place somewhere else—in England, in the United States—through the help of Bohr's personal actions. Then came the years of war; Denmark was occupied by the Nazis in April 1940; Bohr was in close connection with the Danish Resistance. He refused to collaborate with Nazi authorities. Soon he was forced to leave Denmark; he had to escape to Sweden and then came via England to the United States.

Now the fourth period of his life began. He joined a large group of scientists in Los Alamos who, at that time, were working on the exploitation of nuclear energy for war purposes. He did not shy away from this most problematic aspect of scientific activity. He faced it squarely as a necessity, but at the same time it was his idealism, his foresight, and his hope for peace that inspired so many people at that place of war to think about the future and to prepare their minds for the tasks ahead. He believed that, in spite of death and destruction, there is a positive future for this world of men, transformed by scientific knowledge.

At that time we see Bohr actively engaged in a one-man campaign trying to persuade the leading statesmen of the West of the danger and the hope that might come from an atomic bomb. He wanted the men in power to make use of this new momentous achievement for the creation of a more open world in which scientific development should bring East and West together in a common endeavor. He saw Roosevelt and Churchill and other important men, and he learned quickly the difficulties and pitfalls of diplomatic life. Although he was quite able to convince a number of important statesmen, including Roosevelt, of his ideas, his meeting with Churchill turned out to be a complete failure. Churchill wanted none of any sharing of secrets with Russia and went even so far as to accuse Bohr of being too friendly with the Russians.

Bohr's great political concept did not come to any fruition. Neither did other attempts of raising the nuclear technology to an international level in order to avoid a nuclear armament race between powerful nations. Shortly after the war, an International Atomic Energy Commission was convened at the United Nations under the chairmanship of Bohr's old friend and collaborator H. Kramers, but grim short-range realities of the East-West conflict prevented any far-reaching step to overcome them and to face the much more serious long-range realities of a nuclear war. Bohr and all others who thought like him—there were a good number all over the world—were deeply disappointed. Bohr ended his efforts for an international understanding on nuclear weapons with his famous letter to the United Nations, written in 1950, in which he laid down his thoughts about the necessity of an open world.

In the last decade of his life, Bohr spent much time in the organization of international activities in science. He participated actively in the founding of the Scandinavian Institute of Atomic Physics (NORDITA) and of a new European Laboratory in which all European countries should participate for the most modern fundamental research in physics. He helped to build the European Center of Nuclear Research (CERN) at Geneva. It houses one of the world's largest particle accelerators, a symbol of Europe's renaissance in fundamental science, after the United States had taken the lead in this field. Physics became a large enterprise; large numbers of people and large machines were necessary to carry out physical research. The high-energy accelerators made it possible to go beyond the structure of the nucleus and to explore the structure of its constituents, the world within the proton and neutron. Bohr recognized this as a logical continuation of what he and his friends had started. He saw the necessity of physics at a large scale, on an international scale. In no other human endeavor are the narrow limits of nationality or politics more obsolete and out of place than in the search for more knowledge about the universe.

Part 2
Survey Essays

Part 2 contains a collection of survey essays that were written for the purpose of describing recent developments in a field of physics. The first article, "Physics in the Twentieth Century," is the most general of the four; it concerns the development of physics as a whole during our century. Its main thrust, however, is directed toward the physics of the fundamental structure of matter: atomic, nuclear, and subnuclear physics. The equally impressive advances in molecular physics, in the understanding of the solid state of materials science, of gas discharges, and of plasma physics are only lightly touched upon. The article is based upon the introductory talk given at the inaugural session of the European Physical Society, a body that comprises as its members physicists from all European countries and was founded in April 1969, as a symbol of European intellectual unity beyond geographical and political boundaries.

The second essay, "Recent Developments in the Theory of the Electron," surveys the developments in quantum electrodynamics that took place in the late forties and made that part of theoretical physics into one of the most accurate and reliable theories. It is based on the assumption that the electron can be regarded as a true point charge without dimensions. The foundations of the quantum electrodynamics still contain internal inconsistencies stemming from the fact that a true point charge produces an infinite electric field and therefore should have an infinite energy and mass. The ideas described in this article have led to a methodology with which one can make use of the theory in spite of the resulting infinite mass and charge of the electron. One can successfully predict results of any experimental situation in which only electromagnetic interactions are relevant by skirting skillfully the question of the electromagnetic mass (and charge) of the electron or any other charged particle. As Sidney Drell has expressed it, a state of "peaceful coexistence" has been achieved between quantum electrodynamics and the question of the self-energy of charged particles. The mathematical methods of electrodynamics are rather formidable for anybody but the expert; this article tries to give an account of the main ideas and results in a simple way. There is a German word that characterizes the type of explanation used by the author in this and in other articles. It is "*Anschaulich,*" which cannot be translated literally. Its actual meaning may be "plainly visible for the mental eye."

The following two essays deal with nuclear physics, the science of the structure of the nucleus. The first, "The Compound Nucleus," analyzes the concept of the "compound nucleus" which was introduced by Niels Bohr in 1936 in order to understand a number of unexpected phenomena observed in nuclear reactions. The most important of these phenomena were the numerous resonances found first by Fermi and his coworkers in 1934 when he exposed matter to slow neutrons. Bohr's concept was extremely successful at the outset, but its interpretation ran into difficulties when more experimental material was available. In particular the discovery of the nuclear shell structure seemed to contradict the basic assumptions underlying Bohr's concept. This article describes the problem and shows that the concept remains valuable and essential when properly interpreted and used in parallel with the shell model. This paper is written on a somewhat more specialized level than other articles in this book.

The second nuclear article, "Problems of Nuclear Structure," is a survey of the state of our ideas about nuclear structure in 1960. It is based upon the retiring presidential address delivered on February 2, 1961 at a meeting of the American Physical Society. Some important new ideas have been developed in this field of physics since that time, but most of the thoughts described in this article can still be considered as relevant today.

The last essay in this section, "Quantum Theory and Elementary Particles," deals with the physics of elementary particles. In the form presented here it is a synthesis of two articles, one of the same title that appeared in *Science* in 1965 and another that appeared in *Scientific American* in 1968 under the title "The Three Spectroscopies." It sketches the development of our concepts regarding the ultimate structure of matter from atomic theory to our present frontier of knowledge. This frontier is often called "high-energy physics" since one makes use of the highest technically available kinetic energies of particles in order to probe deeper into the structure of matter. Many new "particles" and new phenomena have been found recently at that exciting frontier of science. The article describes some of the attempts to bring order and some understanding into this new wealth of phenomena, but the fundamental insights into the basic laws of material behavior at this level are still to come.

Physics in the Twentieth Century

The spirit of modern science has its roots deep in the culture of antiquity, in Greece, in China, and in the Judeo-Christian tradition, but it came to life and started its exponential growth in the days of the Italian Renaissance. Since then, scientific knowledge and experience have accumulated at a steadily increasing pace until, in our century, the human mind has been challenged to some of its most penetrating insights into the workings of nature. A description of all these accomplishments is much too great a task to be dealt with within the framework of a brief article, and I must restrict myself to sketching this subject with a few short strokes—al fresco, as it were. Reviewing the development of physics in the twentieth century is indeed a dazzling experience. Relativity, quantum theory, atomic physics, molecular physics, the physics of the solid state, nuclear physics, astrophysics, plasma physics, particle physics—all these are children of the twentieth century.

There was a definite change in the character of physics at the turn of the century. The older physics was under the spell of the revelation of two fundamental forces of nature: gravity and electromagnetism. The development of classical mechanics, from Galileo and Newton to Lagrange and Hamilton, had shown that the same natural law, the law of gravity, was operative on earth and in the universe. Electrodynamics, a child of the nineteenth century, reared by Faraday, Maxwell, and Hertz, was the first extensive application of the field concept in physics. The electromagnetic field was recognized as an independent entity in space, and the decisive role of electric phenomena in matter was revealed. The recognition, in the nineteenth century, of the nature of heat as random motion was another of these lucid flashes of rational perception of what is going on in our environment. The development of kinetic theory and thermo-

Revision by the author of an article originally in *Science* **168**, 3934, 923, 22 May 1970.

dynamics in the work of Carnot, Clausius, Helmholtz, Boltzmann, and Gibbs led to thinking in terms of the atomic and molecular structure of matter. The existence of such elementary units was known in the nineteenth century; estimates of their size and weight were not too inaccurate. But the properties of matter were not understood at that time; they were not deduced from more elementary concepts, they were measured and expressed in the form of specific constants of materials, such as elasticity, compressibility, specific heat, viscosity, conductivity of heat and electricity, and dielectric and diamagnetic constants. The books of Lord Rayleigh are perfect examples of a nineteenth-century physicist's view of nature. They give a full and elegant treatment of electric and magnetic fields and of classical wave phenomena of light and sound, together with the properties of solids, liquids, and gases as derived from a number of empirical constants.

The physicists of the nineteenth century were not unaware of the importance of interatomic forces for the determination of material properties. We remember Maxwell's study of the repulsive force between molecules of a gas. But there was no way of telling what the origin of these interatomic forces was, and how to account for their strength or absence, whatever the case may have been. The great variety among the properties of the different elements was not considered a topic for physicists; it was the task of the chemists to analyze and systematize them, as was done so successfully a hundred years ago by Mendeléeff in his periodic system of the elements. The specific features of the different species of atoms—their characteristic optical spectra and their chemical bonds—were known and cataloged by the chemists, but they were not considered a suitable subject for physics. The innate forms and ever-recurring qualities of the atomic species were things foreign to the conceptual structure of nineteenth-century physics. The electron had already been discovered by 1900, and it was obvious that electrons must be essential parts of the atomic structure, but classical physics could not give any clue as to the kind of structure one should expect within atoms.

A Golden Age of Physics

In physics, the twentieth century truly begins in the year 1900. This date is not an accident, it is the year of publication of Max Planck's famous paper on the quantum of action, the birth year of quantum theory. It is impressive to contemplate the rate of progress in physics in the first quarter of this century: Planck's quantum of action in 1900; Einstein's special relativity theory in 1905; Rutherford's discovery of atomic structure in 1911; Van Laue's scattering of x rays from crystals in 1912; Bohr's quantum orbits and explanation of the hydrogen spectrum in 1913; Einstein's general relativity theory in 1916; Rutherford's first nuclear transformation in 1917; Bohr's explanation of the periodic table of the elements (*Aufbauprinzip*) in 1922; the discovery of quantum mechanics by de Broglie, Heisenberg, Schrödinger, and Born in the period 1924 to 1926; Pauli's discovery of the exclusion principle in 1925; Uhlenbeck and Goudsmit's discovery of the electron spin in 1927; Dirac's relativistic quantum mechanics in 1928; Heitler-London's theory of the chemical bond in 1927; and Bloch and Sommerfeld's theory of metallic conductivity in 1930. Let us stop here, although the progress by no means ended in 1930; it went on at this rate for at least another ten years, before slowing down to the relatively slow pace of today.

Among the great systems of ideas which were created in that period, relativity theory—special and general—has a place somewhat different from the others. It was born in the twentieth century as the brainchild of one towering personality. It is a new conceptual framework for the unification of mechanics, electrodynamics, and gravity, which brought with it a new perception of space and time. This framework of ideas is, in some ways, the crown and synthesis of nineteenth-century physics, rather than a break with the classic tradition. Quantum theory, however, was such a break; it was a step into the unknown, into a world of phenomena that did not fit into the web of ideas of nineteenth-century physics. New ways of formulating, new ways of thinking had to be created in order to gain insight into the world of atoms and molecules, with its discrete energy states and characteristic patterns of spectra and chemical bonds.

The new ways of thinking were formulated and codified in the midst of the third decade of this century. The wave-particle duality was proposed by de Broglie in 1924; the equation for particle waves was conceived by Schrödinger in 1925. In these years the concepts of quantum mechanics were expressed and critically analyzed in Copenhagen under the leadership of Niels Bohr, with the help of ideas of Heisenberg, Kramers, Pauli, and Born. The ink of these papers was hardly dry when the new way of thinking began to provide explanations for almost all the atomic phenomena that had been puzzling physicists since they were discovered. The rules of quantization of Bohr and Sommerfeld, which seemed arbitrary when they were proposed, turned out to be logical consequences of quantum mechanics; atomic spectroscopy became a deductive science; Niels Bohr's semiempirical *Aufbauprinzip* emerged logically from quantum mechanics with the help of Pauli's exclusion principle. Mendeléeff's periodic table of atomic properties was easily explained. A few years later, the chemical bond was understood to be a quantum mechanical phenomenon; so was the structure of metals and of crystals. A variant of a famous Churchill statement can aptly be applied to this golden age of physics: "Never before have so few done so much in such a short time."

There are three characteristic features that quantum mechanics has brought to our view of the atomic world.

First, it has introduced a characteristic length and energy

which dominate the atomic phenomena, endowing them with a scale and a measure. The combination of electrostatic attraction between the nucleus and the electron, on the one hand, and the typical quantum kinetic energy of a confined electron, on the other, define a length (the Bohr radius) and an energy (the Rydberg unit). The size of the atoms is determined by that length, which is the combination h^2/me^2 of a few fundamental constants, where e is the unit of charge, m is the electron mass, and h is the quantum of action. The Rydberg unit is given by the combination me^4/h^2. Thus atomic sizes and energies are basically determined and explained.

Second, quantum mechanics has introduced a "morphic" trait previously absent in physics. The electron wave functions represent special forms or patterns of simple symmetry, characteristic of the symmetry of the situation that the electron faces in the attractive field of the nucleus and of the other electrons. These patterns are the fundamental shapes of which all things in our environment are made. They are directly determined from the fields of force that bind the electrons. They always appear, identical and unchanged, whenever the atom finds itself under the same conditions. The appearance of characteristic forms and patterns (Fig. 1) is closely connected with a new way of dealing with mechanical concepts in quantum mechanics. The position of an electron has only probabilistic meaning within a given electron pattern and the same holds for other mechanical attributes such as momentum. This lack of definiteness, usually expressed in terms of uncertainty relations, is more than offset by the much more refined description of atomic reality and observation provided by quantum mechanics.

Also a concept of ideal identity has been introduced. Two atoms are either in the same quantum state, in which case they are identical, or in different quantum states, in which case they are definitely nonidentical. The continuous transition between

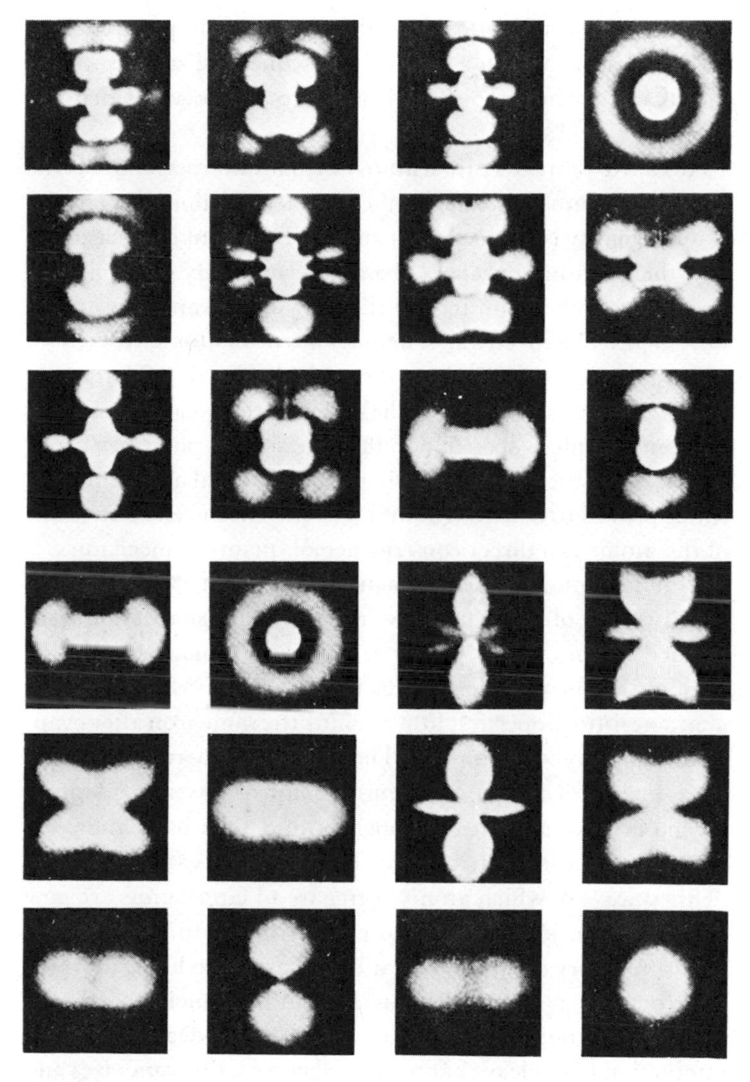

Figure 1.
Patterns of electron distributions in spherically symmetric fields. These are photographs of models.

identical, almost identical, and different has disappeared. The identity has measurable physical consequences, such as the intensity change in the spectra of molecules composed of identical atoms.

The third characteristic feature of quantum mechanics is the use of quantum numbers for the characterization of quantum states. Quality is reduced to quantity: The number of electrons and the quantum numbers of a given state fully determine all properties of the atom in that state. Pythagorean ideas are reborn here: The spectrum of frequencies of an atom represents a characteristic series of values, the typical "chord" of that atom, as it were; the "harmonies of the spheres" reappear in the world of atoms. Kepler's speculation that the sizes of planetary orbits in the solar system constitute simple geometrical and numerical ratios proved to be wrong, but it is reborn in the electron orbits of the atom, as a direct consequence of quantum mechanics.

A fundamental problem of natural philosophy was solved by the discovery of laws which give rise to specific shapes and well-defined entities. Clearly, nature is basically made up of such entities, as our experience tells us every day. Materials have characteristic properties, iron remains the same iron after evaporation and recondensation. The specific properties of matter were first the concern of chemistry, not of physics; quantum mechanics explains these properties and thus has eliminated chemistry as a separate science. The infinitely varied but well-defined ways in which atoms aggregate to larger units are now within range of a rational interpretation in quantum mechanical terms. A theory of the molecular bond came into being in which electron wave patterns (orbitals) keep atomic nuclei together in the right arrangement. Since here one is again dealing with the interaction of nuclear charges and electrons, the same sizes and energies must appear as in atoms—sizes and energies that give rise to interatomic distances of a few Bohr radii and bonding energies of the order of electron volts.

Atomic aggregates consist of two kinds of particles—heavy nuclei and light electrons—which are bound to each other by mutual attraction. The interatomic distances are fixed by the size of the electron cloud, a dimension which can be regarded as the amplitude of the zero-point oscillation of the electron. Because of the much greater mass, the zero-point oscillations of the nuclei in a molecule are much smaller than those of the electrons (the ratio is the square root of the mass ratio); hence in molecules and solids the nuclei form a rather well-localized skeleton, whereas the electrons are distributed continuously like the soft parts of a body around the skeleton. The well-localized arrangement of the nuclei within the molecules introduces the characteristic structural features into chemistry and materials science, with all its architectonic consequences.

The quantum mechanical description of atomic aggregates leads to an understanding of all the material properties and material constants on which classical physics had collected empirical information. In principle, all the constants mentioned above can be predicted and expressed in terms of the fundamental constants e, m, and h and the nuclear masses. For example, the resistance against compression that characterizes solid matter comes from the fact that a reduction in volume leads to an increase in the quantum kinetic energy of the electrons. This is what replaces the "hardness" of atoms in the classical frame of thought.

The well-defined structure of the nuclear framework in molecules is of special significance in macromolecules, which are long linear arrays of molecular groups. The numerous possibilities of different orderings of these groups, each being well defined and reasonably stable, is reflected in the numerous species of living systems in our flora and fauna and is due to an intricate copying and reproduction process, which has been unraveled during the last decade. So chemistry, materials science, and molecular biology are direct descendants of the quantum mechanics of electrons in the Coulomb field of atomic nuclei. The basic structures have a limited stability measurable in fractions of the characteristic energy unit, the Rydberg. Perturbations of a strength of a few electron volts would disrupt them. This is the tender world of chemistry and biology which is destroyed at temperatures corresponding to particle energies of more than a few electron volts, such as exist in most stars. Matter in the form in which we are accustomed to see it is a rare phenomenon in the universe.

A New World of Phenomena

The faint glow of radium in Madame Curie's hand was a telling indication of the existence of yet unknown phenomena in matter. It was apparent from radioactive processes that there must be energies much higher than a Rydberg unit within the atom. Rutherford made use of these processes to penetrate the structure of atoms, and, from anomalous scattering of alpha rays in atoms, discovered the atomic nucleus. Incredible as it may seem, it was only 6 years later (in 1917) that he used the same tool to study the composition of the nucleus and found that some of the constituents are protons. A new world of phenomena had been discovered. However, it was not until 15 years later, in the great year of physics, 1932, that the composition of the nucleus was disclosed. In that year Chadwick discovered the neutron, Fermi published his theory of radioactive beta decay, and Anderson and Neddermeyer discovered the positron. Each of these discoveries had far-reaching significance.

The existence of the positron demonstrated the validity and depth of Dirac's relativistic wave equation (1927), one of the most remarkable examples of the power of mathematical thinking. This equation—a marriage of quantum mechanics with relativity theory—demonstrated the necessity of the existence of an electron spin with its typical magnetic moment. In addition, the equation exhibits a fundamental symmetry corresponding to the existence of two types of matter, ordinary matter and antimatter, with equal properties but opposite charges and other characteristic quantum numbers. Matter and antimatter can be created in empty space if energy is available, and can be caused to revert to pure energy in the reverse process of annihilation. These unusual features had been anticipated theoretically from Dirac's equation before they were discovered in nature.

The discovery of the neutron as a constituent of the nucleus revealed the existence of a new force of nature. It pointed toward a strong nonelectric effect which keeps neutrons and protons tightly bound within the confines of the nucleus. Here was a manifestation of something new—a new force of nature without any analog in macroscopic physics. The "strong interactions" had been discovered.

Fermi's theory of the beta decay demonstrated the existence of another interaction between elementary particles. A neutron can transform itself into a proton with emission of a lepton pair—an electron and a neutrino. This transformation is effected by the so-called weak interaction—a fourth interaction supplementing gravitational, electromagnetic, and strong interactions. It is so weak that the time scale of its nuclear processes is of the order of seconds, days, or years.

Thus the year 1932 was the beginning of a new type of physics dealing with the structure of the nucleus and with its constituents, and working with hitherto unknown forces and interactions.

Let us return to the force between neutron and proton. Scattering experiments have revealed that this force has a rather complicated structure. It is short-ranged and attractive, except for small distances of less than a Fermi, when it becomes repulsive (see Fig. 2). Also it is strongly dependent on the relative spin orientation of the two particles and on the symmetry of the wave function. In this respect, and in its repulsive nature at small dis-

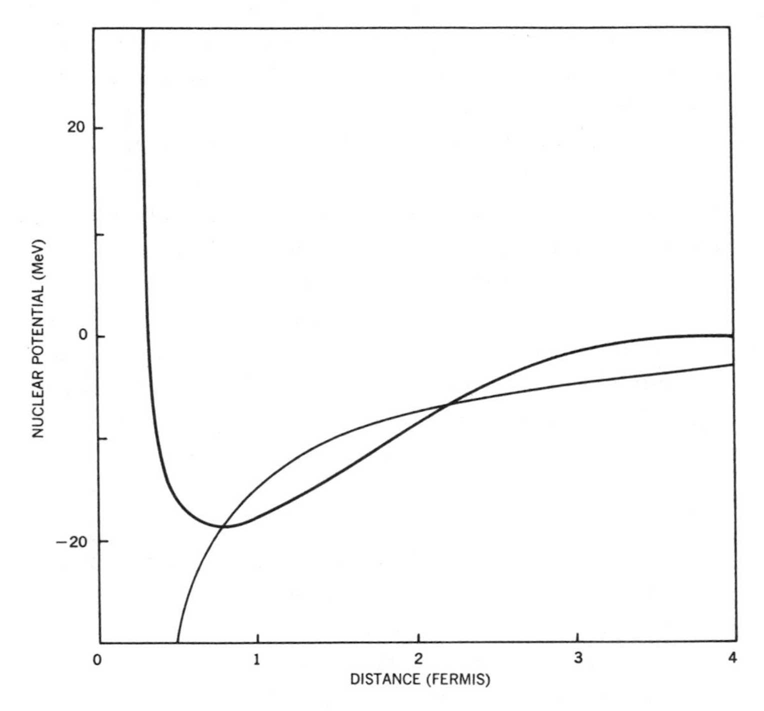

Figure 2.
The potential of the force between two nucleons. The solid curve is an approximate rendition of this potential. The exact value depends on the relative spin direction of the nucleons and on the symmetry of the quantum state. The dashed curve indicates the electrostatic potential between two particles of opposite charge, 3.3 times the charge of the electron. [From *Scientific American,* with permission.]

tances, it resembles the chemical force between two atoms, an analogy to which I return later. In estimating the strength of the attraction, let us compare it with the electrostatic attraction which would be present if the neutron and the proton had opposite electric charges g and $-g$. Of course, the nuclear attraction is short-range and changes to repulsion at small distances, but a qualitative comparison between the electrostatic attraction and the attractive part of the nuclear force is useful. It turns out that the nuclear attraction is roughly equivalent to an electric attraction between two opposite charges of magnitude $g \simeq 3e$. This information allows us to estimate the approximate size and energy of simple nuclear systems by applying the same quantum mechanical principles that were applied in the case of the atom. All we have to do is take the expressions for the Bohr radius and the Rydberg unit, replace e by g, and substitute the nuclear mass for the electron mass. We then obtain the nuclear Bohr radius $a_N = h^2/mg^2 \simeq 2 \times 10^{-13}$ cm, and the nuclear Rydberg $Ry_N = mg^4/2h^2 \simeq 3$ Mev. Nuclear systems are 10^{-5} times smaller than atomic systems, and the relevant energies are in the million-electron-volt region.

Once the nuclear force was established, quantum mechanics could be applied to the nucleus as a system of neutrons and protons. We find in nuclear physics a repeat performance of atomic quantum mechanics, but with a different scale of units. Nuclear-energy-level spectra presented a structure similar to that of atomic spectra, with the same kind of quantum numbers. One significant addition appeared, however—the isotopic spin quantum number. It originates from the fact that the nuclear force does not distinguish neutrons from protons, so that, for nuclear conditions, one should consider the two particles to be two equivalent states of a single particle, the nucleon. Thus a situation arises that is formally similar to the two ordinary spin states of a fermion, and this analogy led Heisenberg to introduce the important concept of isotopic spin and its quantum numbers. The weak interactions provide a process for changing a neutron into a proton and vice versa, so that the spin analogy has also a dynamical sense. The nuclear system therefore is not an entity with fixed numbers of neutrons and protons; all that is fixed is the total number A of nucleons. All nuclei with equal A belong to the same quantum system, and one finds many typical similarities between quantum states of nuclei with equal A which differ only with respect to the number of neutrons that have been replaced by protons. Whereas transitions between atomic states are accompanied by the emission and absorption of light quanta (or by equivalent processes), nuclear transitions are accompanied not only by light radiation but also by weak interactions with emission of lepton pairs, in which case the charge of the system is no longer fixed but changes by one unit.

There are many striking parallels in atomic and nuclear structure. One is the periodicity of properties as a function of the atomic number A, arising from similarity of shell structure. The role of the noble gases, which have high stability and low reaction rates, is played in nuclear physics by those nuclei for which shells are completed. The occupation numbers at which the shells are completed are different, however, because of differences in the average potential and because of the important role that spin orbit coupling plays in nuclei. There exists a Mendeléeff table of periodic nuclear properties too. It is interesting to compare the dependence of certain properties on the number of protons in atoms and in nuclei—properties such as excitation and binding energies, or atomic volumes and nuclear quadrupole moments. Both, atoms and nuclei, show the same kind of periodicity, and the influence of shell structure is manifest.

A nuclear chemistry analogous to atomic chemistry exists, but there is an essential difference. In atoms and molecules, some of the constituents, the atomic nuclei, are well localized; they stay apart from each other and form the skeleton of molecules. This is not the case in nuclear structure. There each constituent is distributed over the whole nuclear volume. Hence, if two nuclei react with each other in a collision, they merge completely. Two oxygen nuclei form a sulfur nucleus and not an O_2 molecule. There is no particle whose zero-point oscillation is small as compared to the object; hence we do not find the variety of forms and the complexity of phenomena that we find in atomic chemistry. Furthermore, there is a limitation on the number of nucleons that can be merged into one unit, because of the electrostatic repulsion of protons.

The Internal Structure of the Nucleon

The analogy between atoms and nuclei is perhaps not thoroughly justified. It is probably more correct to compare nuclei with molecules, where the nucleons play the role of the atoms. Why? The force between nucleons is complicated, in its dependence both on the distance and on other properties. That force is much more like the chemical force between atoms, with its repulsive character at small distances, its minimum of potential in between, and its dependence on the symmetry of the wave function. It is tempting to assume that, in analogy to the chemical force, the nuclear force is not a fundamental force such as the electrostatic attraction; it may be a derived effect of a more basic phenomenon residing within the nucleon, a residue of something much simpler and more powerful, just as the chemical force is a residue of the Coulomb attraction between electrons and nuclei within the atom.

Modern particle physics has discovered much evidence for an internal structure of the nucleon, but it has not yet been able to interpret it. The most important evidence is the fact that the nucleon seemingly changes its character when it is bombarded with beams of energetic particles. It can be excited to a large number of quantum states. These states form a level spectrum which represents a third spectroscopy in which excitation energies are measured in billions of electron volts, not in millions of electron volts as in nuclei, or in electron volts as in atoms. This level spectrum shows regularities similar to those of the other spectra, and the same quantum numbers appear, plus a new one introduced by Gell-Mann and Nishijima, the hypercharge or strangeness. Here again transitions between the states occur with emission or absorption of light quanta and lepton pairs, but a new form of energy exchange was found: the absorption or emission of mesons.

The analogy between nuclei and molecules is enhanced by the character of the spectrum of nuclei, if one considers not only the excitations of the proton-neutron system but also the internal excitations of the nucleons within the nucleus. One obtains a spectrum in which the former excitations are added to each

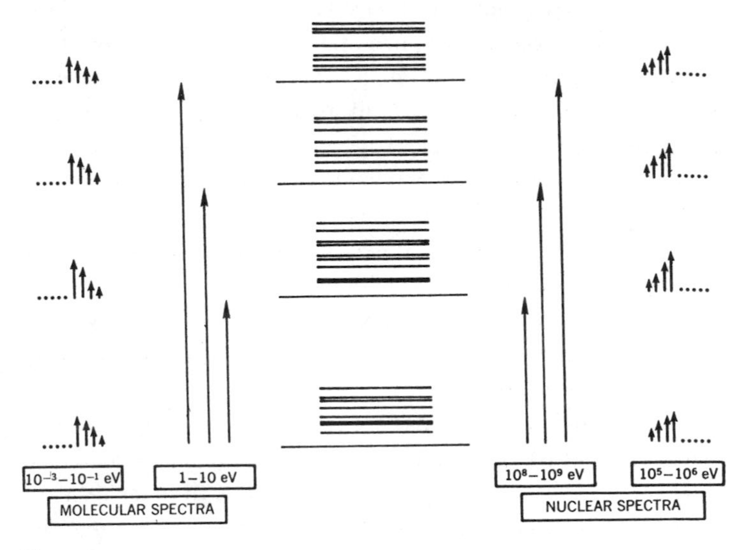

Figure 3.
Schematic sketch of a molecular spectrum and of a nuclear spectrum. The arrows at left indicate the nature and energies of the levels in molecules; those at right refer to nuclei.

internal nucleon excitation. The spectra of hypernuclei, as it were, are included in the nuclear spectrum. This spectrum is strongly reminiscent of molecular spectra (see Fig. 3), in which the rotational-vibrational structure is added to the electronic excitations. There is a quantitative difference, however. In a certain sense the nuclear force is less effective than the chemical force, as shown by the fact that the binding of the deuteron is so weak that it would dissociate even if it were rotating with one quantum of angular momentum. The nuclear force is barely strong enough to concentrate the wave function of the deuteron in the ground state sufficiently within the range of the force for binding to ensue. The bonds of diatomic molecules, however, are able to withstand the centrifugal force of 20 to 40 units of angular momentum. The same contrast is seen in the fact that the binding energy of a nucleon within the nucleus is much smaller than its internal excitation energies, whereas in molecules these two energies—molecular binding and atomic excitation—are comparable. Probably it is more appropriate to compare the nuclear force with the van der Waals force between closed-shell atoms. Nuclear matter would then correspond to superfluid helium, an analogy which goes surprisingly far in explaining the relatively independent motion of nucleons within the nucleus (the shell model) and some typical properties of the spectra connected with a phenomenon corresponding to the energy gap of a superfluid.

Let us go back to the search for the internal structure of the nucleon. This is one of the most challenging frontiers of modern physics. The problem of the nature and structure of the nucleon must be solved if we are to obtain answers to many fundamental questions regarding the basic reasons why matter has the properties we observe and regarding the origin of matter in the history of the universe. Most probably, new insights will be obtained, by this research, into the meaning of commonly used concepts such as particle, field, mass, interaction, and charge. The task is difficult on the experimental as well as the theoretical front, and progress is slow. An encouraging example, however, of the tremendous strides physics has made during this century in penetrating the structure of matter is the development of experiments on inelastic electron scattering.

Franck and Hertz in 1914 demonstrated the existence of excited states in atoms in their famous experiment which showed that the energy losses of scattered electrons equal the different excitation energies. Fig. 4 shows these peaks at those specific energies for electrons scattered from helium atoms. The energy

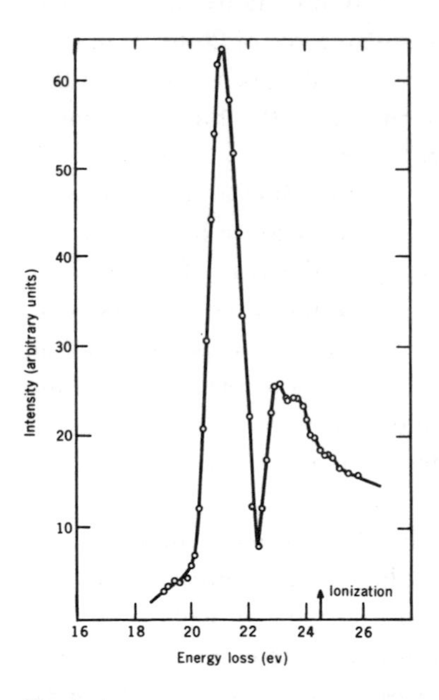

Figure 4.
Intensity of inelastically scattered electrons (energy, 120 electron volts) from helium atoms, as a function of the energy loss. The maxima show the first excited states of helium. [From Mott and Massey, *Theory of Atomic Collisions* (Oxford University Press, New York, 1952).]

losses are of the order of electron volts. Fig. 5 shows results of the same experiment performed with nuclei; one finds the same characteristic peaks, but here we are dealing with millions of electron volts. Fig. 6 shows the results of a recent experiment on inelastic scattering of 16-Gev electrons at the proton; here the peaks in the energy loss correspond to the excitation of the proton itself, which is of the order of several hundred million electron volts—the same type of phenomenon as that observed by Franck and Hertz, but in an energy range 10^8 times larger!

The existence of excited states certainly points to some internal

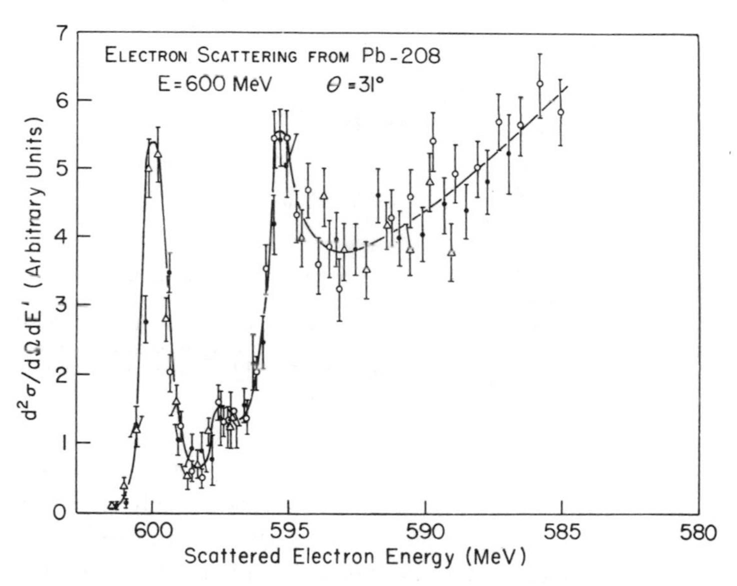

Figure 5.
Intensity of inelastically scattered electrons (energy, 600 Mev) from lead nuclei (Pb^{208}), as a function of the energy of the scattered electrons. The quantity plotted is the differential cross section per unit energy and solid angle as a scattering angle of 31°. The first peak shows the elastically scattered electrons, the next peaks show excitations of the nucleus. [From experiments by H. Kendall and J. Friedmann at Stanford Linear Accelerator Center, Stanford, California.]

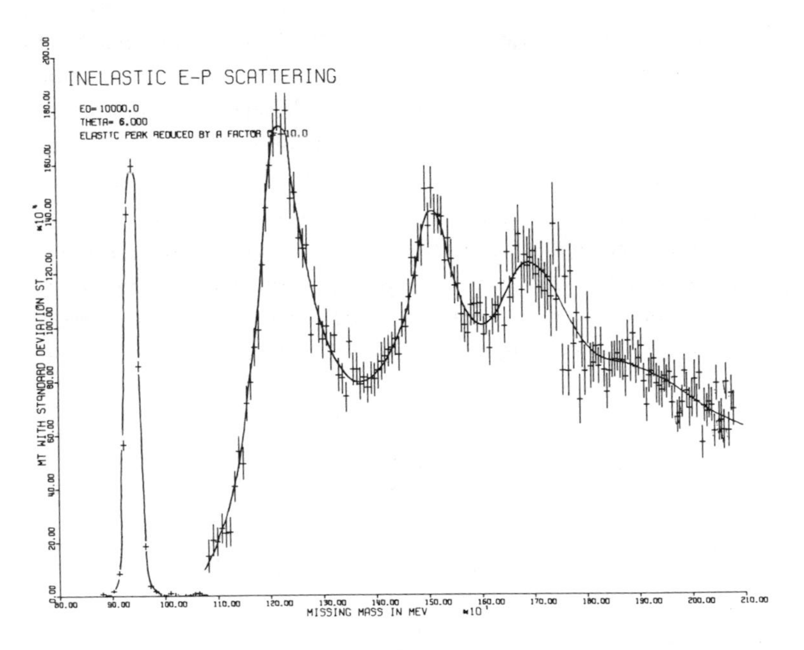

Figure 6.
Intensity of inelastically scattered electrons (energy, 16 Gev) from the protons, as a function of the internal energy of the scattered protons. The quantity plotted is the same as in Figure 4; the scattering angle is 6°. The first peak shows the elastic scattering (reduced by a factor of 10); the next peaks are excitations of the proton. [From experiments by H. Kendall and J. Friedmann at Stanford Linear Accelerator Center, Stanford, California.]

dynamics of the nucleon. The presently known spectrum of these states exhibits certain regularities which are vaguely related to those of a system consisting of three particles with half-integer spin, sometimes referred to as "quarks," "dions," or "stratons." One is tempted, therefore, to consider the nucleon to be made up of three of those subparticles. In addition, the regularities in the spectrum of mesons—they also have been observed in many quantum states that form a spectrum—point toward the hypothesis that a meson is made up of a quark and its antiparticle. In this hypothetical picture the mesons are the quantum states of a quark-antiquark system, in analogy to positronium, which is an electron-positron system. Such systems are forms of pure energy; they can be created by, and annihilated into, other forms of energy. Hence it is a suitable picture for a meson which is absorbed or emitted in transitions between excited states of the nucleon.

Quarks have not been observed in nature. If they really existed they would have some unusual properties, such as an electric charge one-third or two-thirds that of the electron unit, and probably some unusual statistics, different from the expected Fermi statistics. All we can say today is that certain observations can be explained by assuming their existence. They would be bound together by an extremely strong, yet unknown, force, which might turn out to be the fundamental strong interaction, whose residual effects give rise to the nuclear forces. If they did exist, the physics of nucleons and mesons would be a third application of quantum mechanics on an even higher energy scale, after successful applications to atomic and nuclear problems. However, the situation would be different in many respects; our present view of quarks as constituents is probably an inadmissible use of an oversimplified picture. There is one important difference between the world of atomic and nuclear systems on the one hand and of excited nucleons and mesons on the other. In atoms and nuclei the excitation energies are very small as compared to the mass energy of the constituents; hence, the existence of antiparticles is irrelevant to the structure of these systems. In the world of nucleons and mesons, however, the excitation energies are comparable to, and sometimes larger than, the rest mass of the system. The relevant energies are of the order of the mass energies of the particles involved. As a consequence, pairs of particles and antiparticles are present within the system and contribute importantly to its structure. Hence the number of constituents of such systems may be quite undetermined; this is a new and essential feature of these systems. Today we have no systematic way of dealing with such situations.

There is also the question of an internal structure of the electron itself, which has not yet been faced. The most puzzling aspect of this question is the existence of the heavy electron or muon, a particle which, seemingly, is in every respect identical with the electron except for its mass, which is 200 times greater. The muon has a finite lifetime; it changes into a second kind of neutrino different from the ordinary one, with the emission of a lepton pair (an electron plus an ordinary neutrino). It may be that the four known leptons—electron, muon, and two types of neutrinos—represent the beginning of a more complicated lepton spectrum. Although today the electromagnetic properties of the electron are extremely well described by the almost perfect theory of quantum electrodynamics, there remain grave questions regarding the nature of the electron: the reason for the apparent uniqueness of the elementary charge; the existence of the heavy electron; the source of the electron mass; and, last but not least, the nature of weak interactions, with their puzzling violations of established symmetries, such as right- and left-handedness and matter-antimatter symmetry.

Modern particle physics has led to the discovery of many unexpected phenomena. Theoretical understanding does not yet go very far, although theoretical physicists have contributed many ideas, models, and analogies in order to correlate and systematize the wealth of experimental material. There is as yet no Rutherford of particle physics, and no Niels Bohr. The lack of success is not due to any lack of intellectual effort, but the great insight into what goes on within a so-called elementary particle is not yet in hand.

"Extensive" Developments

So far I have sketched the development of our knowledge of the structure of matter in the twentieth century, from atomic physics to modern particle physics. It is not only in this intensive direction toward smaller sizes, higher energies, and phenomena and laws hidden deep within the units of matter that science develops. There is also an "extensive" direction of development, in which knowledge of the basic laws and properties of matter is applied to the understanding of broader fields of inquiry.

Much has been learned and understood since the great breakthrough of quantum mechanics in the third decade of this century. An enormous amount of new insight has been gained into the properties of matter in its varied forms and states of aggregation. The amount of new knowledge is so great that I cannot hope to do justice to it within the frame of this article, which emphasizes fundamental research. I will restrict myself to a few scattered examples. Modern solid-state physics can give a detailed account of the behavior of metals, semiconductors, and crystals of all kinds. In particular, the behavior of solid matter at very low temperature revealed phenomena, such as superconductivity, which, for a long time, defied all explanation. But these phenomena, together with the superfluidity of certain liquids at low temperature, turned out to be understandable and derivable from the basic assumptions about the quantum nature of atomic dynamics. It was a long time before adequate concepts were found which made it possible to formulate the main features of the quantum behavior of systems with many constituents. Once such concepts were formed, they contributed to an understanding not only of the strange behavior of bulk matter at low temperatures but also of some features of the behavior of heavy nuclei, and they helped even in elucidating some problems of quantum electrodynamics and other field theories. The applicability of new concepts in many fields of physics is one of the gratifying developments which emphasize the unity of physics. This is even more apparent in the development of new instrumental methods in experimental physics. The great progress of microwave techniques has advanced all fields, from solid-state physics to elementary-particle physics. The development of our knowledge of materials such as semiconductors gave rise to new and improved particle-detection devices, and beams of elementary particles are the finest tools for the study of interatomic fields in liquids and solids. The strongly coherent light beams which are produced today in lasers and masers have their uses in all fields of physics.

New vacuum techniques, microwave devices, and strong magnetic fields made it possible to study matter in its plasma form—that is, a form of matter at high temperatures and low pressure where most electrons are no longer in their atomic quantum orbits. This state of matter is very common in the universe, in the interior of stars as well as in the expanses of space. The behavior of the plasma state is defined by very simple laws: the electromagnetic interaction between nuclei and electrons. Quantum effects are negligible because of the high excitation. Hence we are dealing with the classical physics of electrons and nuclei. Surprisingly, the resulting phenomena are more complex than those in quantum physics. Superposition principle and quantum stability are absent, and we face strong nonlinear effects and many instabilities.

Modern Astrophysics

An account of physics in the twentieth century would be very incomplete without some mention of astrophysics. It is a science born in this century. It is the frontier of physics at extremely large distances, in contrast to particle physics, which is the frontier of extremely small distances. There is good reason to believe that the two are intimately related. Two major insights have shaped this branch of science: (i) recognition that nuclear reactions are the source of stellar energy, and (ii) the discovery of the expanding universe.

The first discovery has shown that nuclear reactions are infinitely more important for the production of energy than ordinary chemical reactions. However, nuclear processes do not occur on earth, except in the case of those few radioactive elements which are the last embers left over from the great supernova explosion in which our terrestrial matter was produced. In order to study nuclear processes, we had to reproduce them in our laboratories. It was no mean feat for man to recreate, on earth, processes which in nature are found only in the center of stars or in big star explosions, and to make technical use of them, even though some of these uses have been destructive ones.

The second discovery, the expansion of the universe, is mysterious but of fundamental significance. A new time and space scale appears. It is the time in which the universe has expanded to its present state, an interval of approximately 10^{10} years. We are very far from knowing what the universe was like at the beginning of this interval, but one fact is sure: The matter of our present universe was in a very different state at that time. The time interval also defines a length (the distance that light travels during that interval); it is the radius of the present universe, from beyond which no message can ever reach us. It defines a maximum size—about 10^{10} light years—in which our world is embedded.

The twentieth century is, for the universe, what the sixteenth century was for the earth, when Magellan's ships sailed around the planet and showed that it has a finite surface. We have learned in this century that there is a finite universe with which we can be in contact, and we almost fathomed its depth when stellar objects were seen with a red shift of the order of unity.

Modern astrophysics has brought a new aspect to physics: the historical perspective. Previously, physics was the science of things as they are; astrophysics deals with the development of stars and galaxies, with the formation of the elements, with the expanding universe. There are many unsolved questions in this history, many phenomena, such as quasars, which are unexplained, but part of the history is fairly well understood. It is the part in which stars are formed from a hydrogen cloud, elements are formed by synthesis from hydrogen, and stars are developing through different states, some ending as cold chunks of solid matter, others ending in tremendous explosions which we observe as supernovas, sometimes leaving behind fast-spinning neutron stars.

One of these explosions occurred in the year A.D. 1054 and left behind the famous Crab Nebula, in which we see the expanding remnants of the explosion with a pulsar in the center. This explosion must have been a very conspicuous phenomenon, in its first days surpassing the planet Venus in brightness. So different from today's attitudes was the mental attitude in Europe at that time that nobody found this phenomenon worth recording. No records whatsoever are found in contemporary European chronicles, whereas the Chinese have left us meticulous quantitative descriptions of the apparition and its steady decline. What a telling demonstration of the tremendous change in European thinking that took place in the Renaissance!

The kinetic energies produced when large stars contract after their nuclear fuel is exhausted are such that individual protons reach energies of the order of several hundred million electron volts, not far from the energy of their rest mass. Therefore high-energy physics will come into play at these stages of development, and all the newly discovered phenomena of nucleon excitation and meson production will take place on a large scale, just as nuclear reactions take place on a large scale in the center of ordinary stars. Perhaps it is significant that such energies are reached when the gravitational energy of a particle becomes of the order of its mass. This is the so-called Schwartzschild limit, at which the gravitational field becomes critically large and the local space is heavily distorted. This may point to a connection between high-energy physics and gravitational phenomena.

The cosmological aspects of matter reveal a certain insignificance of electronic quantum physics in the universe. Only rarely is matter in a state where the quantum properties of electrons around nuclei are of relevance; for the most part, matter is too hot or too dilute. But it is at those special spots where quantum orbits can be formed that nature developed its atoms, its aggregates, its macromolecules, and its living objects. It is there that the greatest adventure of the universe takes place—that nature in the form of man begins to understand itself.

Recent Developments in the Theory of the Electron

The application of the microwave technique to spectroscopy has greatly increased the accuracy of spectroscopic measurements. Recent experiments on the spectrum of hydrogen and other simple elements have revealed that the results are not exactly in agreement with our fundamental theories of the mechanics of the electron within the atom. Small deviations were found in checking the values of the energy levels in hydrogen given by the Sommerfeld formula [1]. The measured value of the magnetic moment of the electron deviated by about 1 in 1000 from the value given by Dirac's fundamental equation of the electron [2].

These experimental findings led to a reinvestigation of the theory, and especially of its weakest point—the interaction of the electron with radiation. This interaction was treated by a theory named "quantum electrodynamics" which, since its inception by Dirac in 1926, suffered from some internal inconsistencies connected with the old problem of the internal structure of the electron. These inconsistencies make it impossible in this theory to calculate radiation phenomena in a rigorous way.

In the last years, however, some theoretical work has been carried out [3] in an attempt to isolate the unsolved problems and inconsistencies within the theory and to increase the accuracy of the predictions of the theory, in spite of the fact that the structure of the electron and its effects are not understood. The results of this development have been quite successful. The theoretical predictions were in complete agreement with the new experiments. The confidence in the fundamental concepts of "quantum electrodynamics" was greatly enhanced.

This article attempts to present an account of this new development in a form which, I hope, is understandable to the physicist who is not specialized in this field. Only a qualitative and very

Revision by the author of an article originally in *Rev. Mod. Phys.* **21**, 2, 305, April 1949.

incomplete picture of the underlying problems can be given. It seemed advisable not to restrict this report to the newest achievements, but to recapitulate briefly the development of our ideas about the electron beginning with H. A. Lorentz's classical electron theory and including the theory of the positron. The significance of the present problems cannot be evaluated without referring to the most important steps in this development.

I. The Classical Electron Theory

There was hardly any other discovery which led to the understanding of so many and varied phenomena as the discovery of the electron.

Many topics which were thought to be unrelated, like optics, electricity, and chemistry, were understood by the same fundamental mechanism on the basis of the electron theory. It was mainly H. A. Lorentz who brought the classical electron theory into a consistent frame. These were his fundamental assumptions: The electron is an elementary particle with a charge e and a mass m; the motion of the electron is determined by classical mechanics if the force acting on the electron is given by the expression:

$$F = e\mathcal{E} + (e/c)(\mathbf{v} \times \mathcal{H}),$$

where e and $\mathbf{v}$ are the charge and velocity of the electron and $\mathcal{E}$ and $\mathcal{H}$ are the electric and magnetic field strengths. The electromagnetic field in turn is given by the Maxwell equations

$$(1/c)(\partial\mathcal{E}/\partial t) - \text{curl } \mathcal{H} = 4\pi\mathbf{i}, \quad \text{div } \mathcal{E} = 4\pi\rho;$$

$$(1/c)(\partial \mathcal{H}/\partial t) + \text{curl } \mathcal{E} = 0, \quad \text{div } \mathcal{H} = 0.$$

The sources of the field strengths are the charge density ρ and a current density $\mathbf{i}$, which are produced by the electrons.

In most cases it is possible to consider the electron as a point charge. The field created by the electron can then be expressed in a simple manner. We quote only one trivial example: an electron at rest is surrounded by an electric field

$$\mathscr{E} = e/r^2, \tag{1}$$

where r is the distance from the electron. The expressions for the field surrounding an electron in motion are somewhat more complicated.

Some additional assumptions had to be made regarding the conditions under which electrons move in matter: Lorentz assumed that there are several electrons in each atom, which are elastically bound to an equilibrium position and thus are able to perform harmonic vibrations with given frequencies. In electric conductors additional electrons were assumed to move freely about. With these fundamental theoretical tools it was possible to explain a great number of phenomena, as for example, the absorption, scattering, and refraction of light by matter, the Zeeman effect, the optical properties of metals for infrared radiation and many more. In many cases the explanation was only qualitative. Some of the detailed features were not understood. The main assumption of the elastic binding of electrons within atoms was unexplained, especially in view of the planetary structure of the atom. The frequencies of the electron within the atom were neither understood nor determined by the theory.

Lorentz also investigated another fundamental problem: How far is it possible to consider the electron as a point charge? He was forced to make some assumptions about the internal structure of the electron in order to apply the electrodynamic equations within the electron. We quote from his book *The Theory of the Electron*,

While I am speaking so boldly of what goes on in the interior of an electron, as if I had been able to look into these small particles, I fear one will feel inclined to think I had better not try to enter into all these details. My excuse must be, that one can scarcely refrain from doing so, if one wishes to have a perfectly definite system of equations, moreover, as we shall see later on, our experiments can really teach us something about the dimensions of the electrons. In the second place, it may be observed that in those cases in which the internal state of the electrons can make itself felt, speculations like those we have now entered upon, are at all events interesting, be they right or wrong, whereas they are harmless as soon as we may consider the internal state as a matter of little importance.

The main point of interest in the question of the structure of the electron can be formulated very simply today, since the equivalence of mass and energy has become commonplace: The total energy E_{st} of the electrostatic field (1) of the electron is given by

$$E_{st} = (1/8\pi)\int \mathscr{E}^2\, dv,$$

where the integration is extended over the whole space. $\mathscr{E}$ is given by (1) outside of the electron, but (1) is, of course, no longer valid "inside" the electron; it is convenient to assume that the charge of the electron is concentrated on the surface of a sphere with the radius a. In this case E would vanish inside and we would get:

$$E_{st} = (e^2/2)\int_a^\infty (dr/r^2) = e^2/2a. \tag{2}$$

Any other assumption as to the charge distribution does not change the general character of (2): The energy of the electric field depends critically upon the radius of the electron. It necessarily contributes to the mass m of the electron, and we obtain from the Einstein relation

$$m = m_0 + (E_{st}/c^2) = m_0 + (e^2/2c^2 a),$$

where m_0 is the "mechanical" mass of the electron, by which we understand all contributions to the mass which are not of electromagnetic origin. Since the total mass m is known experimentally, there is a lower limit for the radius, corresponding to the assumption that all the mass is of electric origin (we exclude the rather artificial choice of a negative value for m_0):

$$a \geq e^2/2mc^2 = r_0. \tag{3}$$

The electron radius is at least as large as r_0, which is usually called "the classical electron radius." Thus we are forced to abandon the notion of an exact point charge.

Lorentz, Abraham, and Poincaré studied at length the consequences of this new picture. It is not of very great interest to discuss the detailed consequences of the assumption of a finite classical electron. Later developments have brought into the picture new features which completely overshadow these classical considerations.

One point should be mentioned, however. At what energy should one expect in the classical theory the radius of the electron to change significantly the results expected with a point electron? It is easy to see that scattering cross sections of electrons by electrons, or of electrons by equal charges (as protons), should be influenced if the energy is high enough, so that the particles could approach to distances smaller than a. This would happen at energies larger than $2mc^2$, that is, larger than one Mev. One may remark that the physicists of that time would have been very much surprised if they had been able to perform these experiments. Instead of finding an effect of the finite extension of the electron, they would have observed the creation of electron-positron pairs. The fundamental connection of the pair creation with the problem of the structure of the electron will be discussed later in this article.

II. The Quantum Theory of the Electron

The problems of the structure of the electron were soon removed from the focus of interest by the successful development of the quantum theory of the electron. The discovery of the quantum of action, Bohr's theory of the quantum orbits in the atom, and the duality of wave and particle properties of the electron led eventually to the development of quantum mechanics. A new interpretation of the mechanical concepts of momentum, energy, position, and velocity was introduced to describe consistently the facts that appear to be contradictory as, for example, the wave and particle properties of the electron, or the stability against collisions of planetary orbits in atoms. The new theory is best known in the mathematical form of the Schrödinger wave equation.

The success of quantum mechanics was overwhelming. Many unsolved problems of classical electron theory were solved. One can now understand and calculate the resonance frequencies of atoms, the stability of electron orbits, and many other facts which cannot be explained in classical mechanics. There is scarcely any phenomenon within the realm of atoms and molecules which, at least in principle, cannot be accounted for by quantum-mechanical description. It is worthwhile to point out that the quantum theory of the electron could explain all forces between atoms, molecules, and electrons as purely electromagnetic phenomena.

Quantum mechanics can answer all questions as to the behavior of the electron (or other particles) in electromagnetic fields, if these fields are given as functions of space and time. Most of the problems in atomic physics can be put into this form by asking: How does the electron move in an electric or magnetic field of a well-defined character? Difficulties do arise, however, if the question, "What fields are created by the moving charges themselves?" is asked. For example, it could not be explained by the theory in this stage that an atom in its ground state does not radiate light, in spite of the fact that charges are in rapid motion.

Nevertheless, it was possible to construct a number of unambiguous rules to calculate the radiation of atomic systems. This was done by means of two principles. One is the *light quantum hypothesis:* Light of frequency ν can only be emitted and absorbed in quanta of an energy $h\nu$.[1] Thus its emission or absorption must be accompanied by a transition from one quantum state to another, whose energy difference is $h\nu$. The other is the *correspondence principle:* Quantum states of very high excitation show the same mechanical properties as one would obtain from a classical calculation of the same problem. Their radiation should then also be equal to the one which is calculated classically. It was possible to derive rules from these two principles with which one could calculate successfully emission, absorption, and scattering of light by atomic systems. If the wavelength of light is large compared to the dimensions, a system in a quantum state n is, in many respects, equivalent to an assembly of classical electric oscillators with frequencies given by

$$h\nu_{nk} = (E_n - E_k),$$

[1]Here, and in what follows, we understand by ν the frequency in 2π seconds, and by h the magnitude usually referred to as $\hbar$. $h = 1.04 \times 10^{-27}$ g cm^2 sec.$^{-1}$

where k is some other state of the system. The effective charge ϵ of these oscillators is given by $\epsilon^2 = e^2 f_{nk}$ where f_{nk} is the so-called oscillator strength:

$$f_{nk} = (2m/h)\nu_{nk}\left|\int \psi_n{}^*\mathbf{r}\psi_k \, dv\right|^2, \tag{4}$$

where the integral represents the matrix element of r between the states n and k.

The problem of the structure of the electron does not enter into this theory. The theory admits the construction of an electronic wave packet with arbitrarily small diameter, even smaller than a if only wavelengths smaller than a are used. The difficulty arising from the field created by such a packet did not arise since the creation of fields by quantum-mechanical systems was not yet clearly defined.

It is worthwhile mentioning, however, that it is no longer possible to measure effects of an electron radius $a = r_0$ by having two electrons collide with an energy of the order mc^2. The wavelength corresponding to this energy is h/mc, which is much larger than r_0. Thus it is impossible at that energy to locate the electron better than within h/mc. During a collision their average distance will be h/mc, and they practically never will be within a distance comparable with the radius.

III. The Relativistic Wave Equation and Quantum Electrodynamics

The quantum theory of the electron needed improvement in two directions: It needed a generalization for high energies in conformity with the theory of relativity and it needed a consistent treatment of the interaction of matter with radiation. It was Dirac who initiated both steps. He was able to devise a wave equation for the electron which fulfilled the relativistic requirements. He made use of the fact that the electron has an intrinsic spin moment whose state, much like the polarization of light, can always be described as a superposition of a spin parallel to and opposite a given direction of reference. Thus, the electron wave had to be considered as a "spinor" wave with two components corresponding to the two spin directions. Dirac has shown that, for a relativistic wave equation, one has to introduce two more components which, for low velocities, are very much smaller than the others. An electron wave is fully described by giving all four components. Dirac's relativistic wave equation determines the mechanics of this four-component wave. For low kinetic energies (small compared to mc^2) two of the components become very small; the two large ones are themselves solutions of the nonrelativistic (Schrödinger) wave equations, each of them corresponding to one of the two directions of the spin.

The nonrelativistic theory had to ascribe arbitrarily a magnetic moment μ to the spin, whose value it took from the experimental results. Dirac's relativistic equation contains implicitly an interaction of the spin with a magnetic field. The resulting magnetic moment of the electron $\mu = eh/2mc$ is in almost exact agreement with the experiment.

The relativistic wave equation of the electron exhibits, however, several fundamentally unacceptable features. The equation admits solutions which correspond to states of a particle with negative rest mass. The kinetic energy in these states is negative; the particle moves opposite to the motion in ordinary states. For example, a particle of electronic charge is repelled by the field of a proton. These states are, of course, not realized in nature and the most obvious trouble comes from the fact that their energy is negative and, therefore, below the energy of the actual lowest state with positive rest mass. There should be radiative transitions with the emission of light quanta from the regular states to the irregular ones. No regular state could be stable since there are an infinite number of states of negative energy to which it could go with the emission of a suitable quantum of light.

These states cannot be excluded simply by stating that they do not exist in nature. The regular states alone are not what one calls a complete set of solutions. Physically speaking, if by a certain measurement the electron is put into some arbitrary state, it will very probably be a combination of states containing some of the irregular ones. Especially if an electron is localized in a region smaller than the Compton "wavelength" $\lambda_c = h/mc$, the states of negative mass will be strongly represented.

We now proceed to Dirac's treatment of the radiation. In order to describe in a consistent way the interaction between matter and radiation, it is necessary to "quantize" not only the motion of the material particles, but also the electromagnetic field. We understand by "quantizing" the consistent application of certain rules, which led from classical mechanics to quantum mechanics. It is relatively simple to apply these rules to the electromagnetic field in an empty space. The field can be decomposed into its "Fourier components"; it can be thought as a superposition of monochromatic waves. Each of these waves has dynamical properties very much like those of a harmonic oscillator. Thus the "quantization" of the electromagnetic field is equivalent to the quantization of a set of harmonic oscillators and, hence, the energy in one monochromatic wave can change only by multiples of $h\nu$. Thus electromagnetic energy of a frequency ν must appear always in portions of the size $h\nu$. This is the light quantum hypothesis. A further important consequence is the zero-point fluctuations: A harmonic oscillator in its state of lowest energy still has a finite amplitude of vibration. Applying this to the electromagnetic field, we conclude that even in the state of lowest energy the electromagnetic vibrations in space are not zero. The state of lowest energy is the state in which *no* light quanta are present. Hence, in this state the mean squares of the field strengths do not vanish.

We now give an estimate of the strength of the field fluctuations averaged over a volume V of linear dimensions $a: V = a^3$. The amplitude B of the zero-point oscillation of an oscillator of frequency ν is given by $B \sim (h/2m\nu)^{\frac{1}{2}}$; it corresponds to a vibration with an energy $h\nu/2$. The main contribution to the field fluctuations in the volume a^3 comes from waves of a wavelength $\lambda\!\!\!^{-} = c/\nu \sim a$.[2] The amplitude should correspond to an energy of $h\nu/2$, one-half light quantum. Now $(1/4\pi)\langle \mathscr{E}^2 \rangle_{Av}\, a^3$ is the field energy content in a^3; this must be put equal to $h\nu/2 = hc/2a$, so that we get approximately

$$\mathscr{E}^2{}_{\text{fluct.}} \sim hc/a^4. \tag{5}$$

It is larger, the smaller the volume chosen.

The interaction between light and matter can now be described as an interaction between two quantized systems: the electromagnetic field, on one hand, and the electron in the atoms, on the other. Such interaction can be treated by the current methods of quantum mechanics. The interaction energy is given by the classical expression,

$$\frac{1}{c}\int (\mathbf{i} \cdot \mathbf{A})\, dv,$$

where **i** is the current density in the atom and **A** is the vector potential in the field. The integral is taken over the space. The two variables **i** and **A** are now physical magnitudes, which must be dealt with according to the rules of quantum mechanics.

[2]We use here the term "wavelength" for the length $\lambda\!\!\!^{-}$ which is $1/2\pi$ times the conventional wavelength.

Dirac has shown that by this method absorption, emission, and scattering of light can be calculated and that the result is equal to the one which was obtained by the correspondence principle. The emission of light in a transition from the state n to the state k, for example, in this theory is described in the following way. At a given time, say $t = 0$, the emitting atom is in an excited state n and all electromagnetic vibrations are in their ground states. Because of the interaction, the excitation energy $E_n - E_k$ goes over into one of the vibrations; it must, of course, be a vibration whose frequency fulfills the condition $h\nu = E_n - E_k$. The probability P that after a time t the excitation energy has gone into the field turns out to have an exponential time dependence: $P = 1 - e^{-\Gamma t}$. Γ is then the emission probability per unit time. The value of Γ is given by

$$\Gamma = (2e^2 \nu_{nk}{}^2/3mc^3) f_{nk},$$

in conformity with the probability of radiation of an oscillator with the strength f_{nk}, as defined in (4).

Dirac's quantum electrodynamics gave a more consistent derivation of the results of the correspondence principle, but it also brought about a number of new and serious difficulties. The structure and size of the electron appeared again in the theory. The trouble arose from the interaction with the electron of the zero-point fluctuations of the field. Let us consider a free electron under the influence of an oscillatory field strength $\mathcal{E} = \mathcal{E}^0 e^{i\nu t}$: It performs forced oscillations of frequency ν with a displacement x_ν. The average square $\langle x_\nu{}^2 \rangle_{Av}$ of this displacement and the average square of the velocity $\langle \dot{x}_\nu{}^2 \rangle_{Av}$ of a free electron are given by

$$\langle x_\nu{}^2 \rangle_{Av} = \tfrac{1}{2}(e^2 \mathcal{E}_0{}^2/m^2\nu^4), \quad \langle \dot{x}^2 \rangle_{Av} = \tfrac{1}{2}(e^2 \mathcal{E}_0{}^2/m^2\nu^2). \tag{6}$$

The kinetic energy of the electron in these oscillations is

$$E_\nu = \tfrac{1}{2}m\langle \dot{x}_\nu^2 \rangle_{Av} = e^2\mathscr{E}_0^2\lambda\!\!\!^-{}^2/4mc^2, \tag{6a}$$

where $\lambda\!\!\!^-$ is the wavelength belonging to the frequency ν. Hence, the zero-point oscillations of the field contribute to the electron a certain amount of energy. Let us assume for a moment that the electron is a sphere with a radius a. Then only waves with a wavelength $\lambda\!\!\!^- > a$ will act upon the electron; the ones with $\lambda\!\!\!^- \gg a$ are not very important, so that we are allowed to put in (6a) $\lambda\!\!\!^- = a$. If we then enter the value (5) for $\langle \mathscr{E}_0^2 \rangle_{Av}$ over a volume a^3, we obtain for the energy E_{fl} of the electron due to the zero-point field fluctuations:

$$E_{fl} \sim e^e h/4mca^2. \tag{7}$$

In a more accurate calculation we start with expression (6), which gives the effects on the electron induced by an oscillatory field strength of amplitude $\mathscr{E}_0$ and frequency ν. In order to calculate the value of $\mathscr{E}_0{}^2$ for the zero-point oscillations, we include the electromagnetic field and the electron into a big volume Ω. The zero-point amplitude $\mathscr{E}_0 e^{i\nu t}$ of one proper vibration can be calculated by putting the total energy of the oscillation equal to:

$$(1/8\pi)\int(\mathscr{E}^2 + \mathscr{H}^2)\,dv = (1/8\pi)\mathscr{E}_0{}^2\Omega = h\nu/2; \quad \mathscr{E}_0{}^2 = 4\pi h\nu/\Omega.$$

We use the well-known formula that there are

$$z(\nu)d\nu = \Omega(\nu^2/\pi^2c^3)d\nu$$

proper vibrations in the frequency interval $d\nu$. Since the zero-point oscillations of different frequencies are statistically independent, their contributions to the average square of the displacement and of the velocity add up and we get for the total of these magnitudes:

$$\langle x^2 \rangle_{Av} = \int \langle x_\nu{}^2 \rangle_{Av} z(\nu)\, d\nu = (2e^2 h/\pi m^2 c^3) \int_{\nu 0}^{\infty} (d\nu/\nu), \tag{8}$$

$$\langle \dot{x}^2 \rangle_{Av} = (2e^2 h/\pi m^2 c^3) \int_{\nu 0}^{\infty} \nu\, d\nu. \tag{9}$$

The integrals are extended between a lower limit ν_0 and infinity. The frequency ν_0 depends on the state of binding of the electron. $h\nu_0$ is of the order of the binding energy. If the frequency of the field oscillations falls below the frequency ν_0, the electron can no longer be considered as free and (6) is no longer valid. The resulting effect is equivalent to an omission of the frequencies below ν_0.

Both expressions (8) and (9) lead to infinite results. This is especially troublesome in the case of the velocity square because it gives rise also to an infinite kinetic energy E_{fl} of the electron due to the zero-point fluctuations:

$$E_{fl} = (m/2) \langle \dot{x}^2 \rangle_{Av} = (e^2 h/\pi m c^3) \int_{\nu 0}^{\infty} \nu\, d\nu. \tag{10}$$

This expression contains a quadratically divergent integral. Since this energy is an inseparable part of the total energy of an electron, it must appear as part of its mass energy mc^2. In order to keep the mass finite, one therefore is forced to assume some structural properties of the electron which prevent the interaction with high frequencies of the field. We can do this by introducing an upper limit ν_{max} to the interaction which cuts off the integral in (10) at that limit. The fluctuation energy assumes the form

$$E_{fl} = (e^2h/2\pi mc^3)\nu^2{}_{max}, \tag{7a}$$

and we can determine an upper bound for ν_{max} by setting E_{fl} equal to mc^2:

$$h\nu_{max} \leq (2\pi hc/e^2)^{\frac{1}{2}}mc^2 \approx 15 \text{ Mev}. \tag{7b}$$

This would remove the interaction with an electron at rest of a quantum of an energy >15 Mev, a rather improbable result. The introduction of ν_{max} is equivalent to the assumption of an electron radius $a = c/\nu_{max}$, which shows the equivalence of (7) and (7a). Equation (7b) gives rise to a value of $a \approx (hc/e^2)^{\frac{1}{2}}r_0$, which is larger than the classical limit (3). Thus the fluctuation energy seemingly pushes the electron radius to even greater values than the one which we obtained from the energy of the electrostatic field. It should be noted, however, that in interactions with light of an energy of more than $2mc^2$, the irregular solutions with negative mass play an essential role. Thus the significance of these states will have an essential bearing upon the problem of the self-energy of the electron.

Dirac's two generalizations of quantum mechanics, the relativistic wave equation and the quantum electrodynamics, were very successful in some respects: the explanation of the magnetic moment of the electron, the derivation of the Sommerfeld fine structure formula, and the consistent derivation of the expressions for the absorption, emission, and scattering of light. Two fundamental difficulties were introduced simultaneously:

(1) The existence of states of the electron of negative mass. They cause an instability of a normal bound state by the emission of a quantum of high energy and subsequent transi-

tions into a state of negative mass. Thus, the "normal" states of the electron have a very strong "resonance" interaction with light quanta of high energy.

(2) The quantization of the electromagnetic field introduces infinite fluctuations of the electron. In order to keep their contribution to the energy within the observed mass energy value, the interaction of the electron with light quanta of an energy $h\nu > 137^{\frac{1}{2}}mc^2$ would have to be basically altered. It will be shown in the next section that the positron theory removes the first difficulty and completely changes the aspect of the second.

IV. The Positron Theory

The phenomenon of creation of a positron and an electron by a light quantum introduces a new aspect into the theory of the electron. The fundamental process can be described as follows: A light quantum of an energy larger than $2mc^2$ (1 Mev) can be absorbed by the empty space, in the presence of strong electric fields. The energy is then transformed into a pair consisting of a positive and a negative electron.

Two outstanding facts are shown in this phenomenon: the existence of a positive electron, and the fact that the vacuum has physical properties, which enables it to absorb light and to produce electrons. Hence, the physical description of the vacuum is bound to be more complicated than hitherto and must contain the latent electron pairs which can be created.

It was again Dirac who, turning a vice into a virtue, used the unacceptable states of negative mass for the description of the vacuum. A reinterpretation of these states gives an almost perfect description of the vacuum and the existence of positrons: The states of negative mass correspond in some respects to the states of a particle of opposite charge since they move in opposite directions in any electromagnetic field. They are, however, still unacceptable because of their negative kinetic energy. The reinterpretation which removes this difficulty can be formulated as follows: According to the Pauli exclusion principle, any state can be either occupied by one single electron, or unoccupied. The occupation of a state of energy E_i increases the total energy of the system by the amount E_i; the removal of an electron from the state decreases the total energy by E_i. Dirac's reinterpretation of the states of negative mass consists in the exchange of "occupation" and "removal." We decide to call an occupied state of negative mass "empty" and an empty state "occupied." The transition from "empty" to "occupied" is then connected with an energy change of $-E_i$. Since E_i is negative itself, the energy actually increases by $+|E_i|$. The trouble with the negative energy is thus removed.

The vacuum can then be described formally by assuming that all states of negative mass are occupied by electrons. They are not "actually" occupied, because of our reinterpretation, so that one need not be bothered by the infinite charge density which one would get if all states of negative mass were really occupied. The wave functions which represent the *absence* of positrons are the same functions which would have represented the *presence* of electrons of negative mass. It is a new feature that the "absence" of a particle is described by a wave function. This is, however, an expression of the fact that the vacuum has the physical properties described above; it is filled with latent electrons.

This reinterpretation removes at once the difficulty which the states of negative mass have introduced. Since in the vacuum these states are occupied, no electron in the regular states can jump into them. Thus the regular states are no longer unstable against decay into the irregular ones. They no longer are in "resonance" interaction with arbitrarily high light quanta.

The pair creation is then described as follows: A light quantum produces a transition from an occupied state of negative mass to a state of positive mass. The result is an electron in a state of

positive mass and an unoccupied state of negative mass. The latter must be interpreted as an occupied state of a positron with positive mass. Thus the light quantum has created two particles positive and negative with positive mass.

Such transition can only occur in the presence of external fields. Without those fields energy and momentum cannot be conserved. The transition probability can be calculated and the results reproduce excellently the experimental material. The opposite process is the annihilation of a positive and a negative electron, with the emission of either one quantum in an electric field or of two quanta in the field-free space. It can be described by our picture as the transition of the electron into the "unoccupied" state by which the positron is represented. This transition is accompanied by the emission of light quanta.

The new aspect of the vacuum has a decisive effect upon the problem of the self-energy of the electron. The properties of the vacuum with respect to the electrons are now, in some aspects, analogous to its properties in respect to the electromagnetic field. There exist also zero-point fluctuations of the electric charge and the electric current in the vacuum. These fluctuations are very small when averaged over a volume of a size larger than the Compton "wavelength" $\lambda\!\!\!^{-}_c = h/mc$. They represent the latent electron pairs which, by means of light quanta, could be brought into real existence.

Let us now consider the properties of the "vacuum" in the neighborhood of an actual electron. There will be an interaction between this electron and the latent charges, mainly because of the Pauli exclusion principle. According to this principle, electrons tend to keep distance from one another. Two electrons (of equal spin) do not come nearer than a distance d which is determined by their relative momentum p: $d \sim h/p$. (They must not be in the same cell of the phase space.) The presence of one actual electron in the vacuum introduces some changes in the

"charge distribution" of the vacuum. This charge distribution would be zero on the average if undisturbed. The wave functions which represent the electrons of negative mass are slightly removed from the place of the actual electron. This change of charge distribution, compared to the undisturbed vacuum, appears as an addition to the "actual" electron. This manifests itself in form of a spread in the charge distribution of an electron, since the vacuum electrons are slightly pushed away from the actual electron. The calculation shows that this spread is enough to change the classical electrostatic self-energy to $(e^2/hc)mc^2 \log(\lambda\!\!\!^{-}_c|a)$. Here a is the "radius" of the electron, or, as a better definition, a is a limit of wavelength so that fields with $\lambda\!\!\!^{-} < a$ are no longer assumed to interact with the electron.

The effects of the zero-point field oscillations are even more drastically changed by our new concept of the vacuum. This comes from the fact that the field oscillations also interact with the latent electron pairs in the vacuum. As long as their frequency is much smaller than $2mc^2/h$ (the minimum frequency of pair creation), the "vacuum" is very little influenced and the old calculation (6) of the displacement $\langle x_\nu{}^2\rangle_{Av}$ and the velocity $\langle \dot{x}_\nu{}^2\rangle_{Av}$ are still valid. For frequencies higher than $2mc^2/h$, however, the field oscillations have a strong effect on the latent electron pairs and the induced charge and current fluctuations in the vacuum interfere with the induced fluctuation of the electron itself. This interference is destructive and reduces to some extent values of the induced displacement and velocity. The reduction can be roughly approximated in its main features by a factor $(mc^2/h\nu)^2$ to the expressions (6) for $h\nu > 2mc^2$:

$$\left.\begin{aligned}\langle x_\nu{}^2\rangle_{Av} &= \tfrac{1}{2}(e^2\mathscr{E}_0{}^2/m\nu^4)\,(mc^2/h\nu)^2\\ \langle \dot{x}_\nu{}^2\rangle_{Av} &= \tfrac{1}{2}(e^2\mathscr{E}_0{}^2/m\nu^2)\,(mc^2/h\nu)^2\end{aligned}\right\}\text{for } h\nu > 2mc^2. \qquad (6')$$

This effect is difficult to explain in qualitative language. It is

connected with the Pauli exclusion principle, according to which electrons have a tendency to keep apart from one another. Thus the charge and current fluctuations of the vacuum in the neighborhood of the electron tend to be in opposite phase to the fluctuations of the electron itself and therefore cause the destructive interference.

These effects represent a definite improvement. The average displacement $\langle x^2 \rangle_{Av}$ no longer leads to infinities. The divergent integral in (8) converges now because of the reduced contribution (6′) of the frequencies above $2mc^2/h$, and we obtain

$$\langle x^2 \rangle_{Av} = (2e^2 h/\pi m^2 c^3)\log(fmc^2/h\nu_0), \tag{11}$$

where f is a factor of the order unity, which can be determined if the effect of the higher frequencies is exactly taken into account. The average velocity square (9) is still infinite but the divergence is only logarithmic. We get from (6′):

$$\langle \dot{x}^2 \rangle_{Av} \sim (2e^2 h/\pi m^2 c^3)\left[\int_0^{2mc^2/h} \nu\, d\nu + (mc^2/\hbar)^2 \int_{2mc^2/h}^{\infty} (d\nu/\nu)\right].$$

The fluctuation energy $E_{fl} = (m/2)\langle \dot{x}^2 \rangle_{Av}$ is reduced to $E_{fl} = (e^2/\pi hc)mc^2 \log(fh\nu_{\max}/mc^2)$ where f is a numerical factor and $\nu_{\max}$ the cut off frequency. In order to keep this energy below the total mass energy mc^2 of the electron, it is now sufficient to keep $a = c/\nu_{\max}$ larger than $(h/mc)\exp[-(\hbar c/e^2)]$. This lower limit is very much smaller than any length considered so far. It is no longer necessary to tamper with the interaction of the electron with light quanta of an energy of a few Mev. It is still unsatisfactory, of course, that the limit cannot be chosen to be infinity without obtaining infinite self-energies; thus the internal structure of the electron will appear somewhere in the theory. However, some changes in the interaction between light and matter are certain to occur at very high-energy values where we have good reason to expect the appearance of new phenomena (nuclear or meson type).

So far we have discussed the influence of an actual electron on the vacuum due to the Pauli exclusion principle. There is also an influence, although weaker, in the form of a displacement of the vacuum electrons due to electric interaction. It is easier to describe this effect, not for an actual electron, but for a *proton*, which is embedded in the vacuum. The wave functions of the states of negative mass are all deformed because of the presence of the proton. Since the vacuum is described by the undeformed states, the difference between the deformed and undeformed ones should give rise to an actual charge density. This is called the polarization of the vacuum by an external charge (the proton).

The proton induces a charge density ρ_i in the vacuum. The calculation shows that $\rho_i(\mathbf{r})$ as function of the location $\mathbf{r}$ has the following form:

$$\rho_i(\mathbf{r}) = A\rho_0(\mathbf{r}) + \int G(\mathbf{r}-\mathbf{r}')\rho_0(\mathbf{r}')\,\mathbf{dr}'. \tag{12}$$

Here $\rho_0(\mathbf{r})$ is the external charge density; in our case, ρ_0 is the charge density of the proton. A is a constant and $G(\mathbf{r}-\mathbf{r}')$ is a function of the distance between the points $\mathbf{r}$ and $\mathbf{r}'$. The integral is extended over all points $\mathbf{r}'$. The expression for the induced charge consists of two parts: The first term is exactly proportional to the inducing charge density ρ_0; the second part is an effect at a distance. According to this term a point charge at $r = 0$ (like a proton) would give rise to a charge distribution $G(r)$. $G(r)$ is different from zero and negative only over distances up to the Compton wavelength λ_c. The effect is the same as if the dielectric coefficient of the vacuum was smaller than unity by about 1/137 over a region of the order λ_c. It is important to note that the first part is unobservable in principle. Its effect is undistinguishable from the original charge density ρ_0, since it is always induced by it. What is actually measured in nature as the charge of the proton would not be e, but $(1+A)e$. It thus represents nothing but a renormalization of the charge. The second term only has physical significance.

There is one serious difficulty with this interpretation: The factor A turns out to be logarithmically infinite: $A \sim (e^2/hc) \log(\lambda_c/a)$ if the "cut off" radius a is put equal to zero. This would mean that the external charge ρ_0 of the proton induces a charge in the vacuum at the same place, which changes its value by an infinite amount. It is true that this change is in itself unobservable, since one always observes the total charge, external plus induced, in nature. However, the fact that the induced charge is infinite for $a = 0$ represents a serious difficulty of the theory.

The vacuum is polarized not only by a proton but also by an electron. The situation is somewhat more complicated in this case because of exchange phenomena between the electron and the vacuum electrons. The fact remains, however, that the electron, if considered as a point $(a = 0)$, also induces a charge in the vacuum which adds an infinite contribution to its original charge. Thus the internal structure of the electron is relevant not only for its mass but also for its charge.

One can make these infinite additions finite without changing the second term in (12) by arbitrarily removing the interaction of the field with electrons whose wavelength is smaller than a. Here, as in the self-energy, the infinity comes from the interaction at very high energies, and there is hope that a future theory will change this interaction so that the constant A remains finite and small.

In spite of these difficulties, the theory of the positron can be regarded as a big step forward in our understanding of the electron: By means of Dirac's reinterpretation of the states of negative mass it was possible to explain the new phenomena of pair creation and annihilation and to remove several fundamental difficulties of the Dirac equation:

(1) The radiative transitions from the ordinary states into states of negative mass are removed.

(2) The fluctuation energy is much less sensitive to the structure of the electron because of its logarithmic dependence on the electron radius.

(3) The average square displacement of the electron by the field fluctuations is finite and independent of the radius or the structure.

V. The Experimental Test of Quantum Electrodynamics

The quantization of the electromagnetic field so far has not brought much reward. It is true that it made it possible to derive the expressions for the absorption, emission, and scattering of light, which before were based only upon a recipe contrived by means of the correspondence principle. On the other hand, new difficulties came about, all connected with the zero-point oscillations of the electromagnetic field and their effect on the self-energy of the electron. Quantum electrodynamics has not yet shown its superiority over the correspondence principle. On the contrary, its actual expressions for the electromagnetic phenomena become senseless, since a consistent interpretation of the theory would force us to put the mass m of the electron equal to infinity at all places where it occurs.

Encouraged by some new experiments, which will be discussed later on, a new attempt was made recently to find observable effects, which are directly connected with the new features introduced by quantum electrodynamics. The main theoretical difficulty consisted in the problem of how to separate the infinities of mass and charge from the rest of the theory, in order to obtain results that can be applied to nature. This was done by isolating the expressions for the infinite mass and charge within the theory, in the hope that mass and charge will be made finite by a future improvement. Such procedure is possible since the self-energy terms and the infinite charge come mostly from the interaction with very high-energy light quanta and are, therefore, largely independent of the state of binding of the electron in fields normally occurring in nature. Hence, they can be split off as an additional mass and charge of the electron. This has been shown already for the charge in the last section by discussing expression (12). The separation of the mass term is mathematically much more complicated but can be performed in an analogous way. The relativistic transformation properties of the terms occurring in the calculation proved to be of great importance for finding an unambiguous rule as to what parts of the expression of the self-energy can be considered as a mass term. It was necessary to reformulate quantum electrodynamics so that the relativistic invariance of the theory was more explicit than before. This very laborious task was performed by J. Schwinger and independently by S. Tomanaga.

There is, however, a small part of the self-energy which is not contained in the mass and which is due to the interaction with oscillations of lower frequencies. This part depends on the external conditions and may give rise to a slight shift of energy levels, depending on the conditions of binding, and a slight change in some of the fundamental properties of the electron. It is due mainly to the effect of the displacement x of the electron by the zero-point oscillations, whose square average $\langle x^2 \rangle_{Av}$ turned out to be finite and due entirely to the interaction with lower frequencies. This can be demonstrated by means of quite elementary calculations [4] in a case which corresponds to an actual experiment, namely, the shift of the levels in hydrogen-like atoms.

Let us consider a stationary state n of the electron in a Coulomb field, whose wave function is given by ψ_n. The Coulomb field is described by the potential energy $V(r) = Ze^2/r$, where r is the distance from the nucleus. The average potential energy $\overline{V}$ in the state n can be written in the form

$$\overline{V} = \int V(r)|\psi_n(\mathbf{r})|^2 \, dv, \tag{13}$$

where $|\psi_n(\mathbf{r})|^2$ is the well-known probability of finding the electron at a point $\mathbf{r}$: The integral is extended over the volume. This expression must be changed in view of the existence of the zero-point oscillations. The effect of these oscillations on the electromagnetic mass is already assumed to be contained in the observed electron mass m. There is, however, also an influence on the potential energy, since the electron is forced to oscillate around the position $\mathbf{r}$. It will be shown that this oscillation changes the average value of the potential energy by a small amount. This change gives rise to a shift of the energy levels.

In order to calculate this change we replace $V(\mathbf{r})$ in (13) by $V(\mathbf{r}+\mathbf{x})$, where $\mathbf{x}$ is the zero-point oscillation of the electron. We use a Taylor expansion because of the smallness of $\mathbf{x}$.[3]

$$V(\mathbf{r}+\mathbf{x}) = V(\mathbf{r}) + \operatorname{grad} V \cdot \mathbf{x} + \tfrac{1}{2}\Delta V \cdot (\mathbf{x}^2/3), \tag{14}$$

where ΔV is the Laplace operation on $V: \Delta V = [(\partial^2/\partial x^2) + (\partial^2/\partial y^2) + (\partial^2/\partial z^2)]V$. The second term is zero in the average, since $\mathbf{x}$ is an oscillation. Thus the addition δE_n to the average potential energy of the state n may be written:

$$\delta E_n = \frac{1}{6}\int \Delta V \cdot \langle x^2 \rangle_{Av} |\psi_n(\mathbf{r})|^2 \, dv.$$

[3]The simple form of the third term in (14) comes from the fact that, in the average: $\langle x_x x_y \rangle_{Av} = 0$, $x_x{}^2 = x_y{}^2 = x_z{}^2 = (\mathbf{x})^2/3$.

The Laplacian of the Coulomb potential is proportional to the charge density ρ_0 which produces it: $\Delta V = 4\pi e\rho_0$, where ρ_0 is the charge density of the nucleus, which we approximate by a δ-function:[4] $\rho_0 = Ze\delta(r)$, where Ze is the charge of the nucleus. Hence we obtain for δE_n:

$$\delta E_n = (2\pi/3)Ze^2|\psi_n(0)|^2\langle x^2\rangle_{Av}, \tag{15}$$

where $|\psi_n(0)|^2$ is the intensity of the wave function at the nucleus, and we insert the value (11) which we found for $\langle x^2\rangle_{Av}$ into (15) to calculate the level shift. The frequency ν_0 which occurs in (11) depends on the binding of the electron and is of the order of the Rydberg frequency ν_R for an electron in a hydrogen-like atom. Since ν_0 appears only under a logarithm, its exact value is not of great importance. It has been shown by Bethe [3] that, for a quantum state **n**, ν_0 is given by the formula

$$\log h\nu_0 - \frac{\sum_m |p_{nm}|^2(E_m - E_n)\log|(E_m - E_n)|}{\sum_m |p_{nm}|^2(E_m - E_n).},$$

where E_n is the energy of the state n and the sums are extended over all other quantum states m. p_{nm} is the matrix element of the momentum between the states n and m.

We observe that $|\psi(0)|^2$ vanishes for all states except S states (states with the orbital angular momentum zero), for which the simple relation holds:

$$|\psi_n(0)|^2 = Z^3|\pi l^3 n^3, \tag{16}$$

where $l = h^2|me^2$ is the Bohr radius. Thus the level shift vanishes for states with an angular momentum different from zero.

[4]The δ-function $\delta(r)$ is zero everywhere except at $r = 0$. It is normalized such that the volume integral $\int \delta(r)dv$ is equal to unity.

We finally get the level shift for S states from (15), (11), and (16). It is practical to express it in form of a relative shift by dividing δE_n by the energy E_n of the level which is given by the Balmer relation $E_n = Z^2 me^4/2h^2n^2$:

$$\delta E_n/E_n = (8/3\pi)(e^2/hc)^3(Z^2/n)\log(fmc^2/h\nu_0). \tag{17}$$

The exact calculation for the $2S_{\frac{1}{2}}$ term yields the values $\nu_0 = 18\nu_R, f = 1.3$. Thus the S levels of hydrogen-like atoms should be shifted upwards (δE_n is positive) by small amounts, relative to the values given by the Sommerfeld formula. This is a direct effect of the zero-point oscillation, and its experimental verification constitutes a strong support of quantum electrodynamics.

The polarization of the vacuum by the proton produces also a shift which has to be added to (16). According to the discussions of the last section, the only observable effect is a small polarization around the proton of the extension λ_c. The calculation shows that this causes a level shift $\delta E_n'$:

$$\delta E_n'/E_n = -(8/15\pi)(e^2/hc)^3 Z^2/n. \tag{18}$$

It amounts to only about 1/40 of the shift δE_n.

The most reliable experiment on the level shift was performed by Lamb and Retherford [1] on hydrogen. According to the Sommerfeld formula the $2S_{\frac{1}{2}}$ level and the $2P_{\frac{1}{2}}$ level of the hydrogen atom should coincide in energy, and the $2P_{\frac{3}{2}}$ level should lie 10,000 megacycles higher. Lamb and Retherford have measured the $2S_{\frac{1}{2}}$ level relative to the two other levels and have found that the $2S_{\frac{1}{2}}$ level is shifted upwards by about 1060 *mc*, a value which is in good agreement with the theoretical formula (17). Similar shifts have been found by J. Mack [5] and Kopfermann and Paul [6] in helium. The present measurements are not accurate enough to prove the existence of shifts as small as the one given by (18), caused by the polarization of the vacuum. Future experiments will show whether this additional effect can be considered as real.

Another important result obtained by these methods is the correction to the g factor of the electron. According to Dirac's equation, the magnetic moment of the electron μ_e is equal to $he/2mc$. The ratio between this value and the mechanical moment $\hbar/2$ of the electron is $g(e/2mc)$ with $g = 2$, in contrast to the value of this ratio for orbital motions in which $g = 1$. If the interaction of the electron with the radiation field is properly taken into account, one obtains the result that g is not accurately equal to 2 but $g = 2 + e^2/\pi hc$.

Unfortunately, it is impossible to give a qualitative description of this effect along the lines in which the level shift was explained. The spin of the electron is in itself a phenomenon which is not amenable to a simple pictorial understanding. A way to understand the effect may be found by remembering that the magnetic moment of the Dirac electron is due to circular currents of the radius h/mc. The zero-point oscillations of the electromagnetic field influence these currents to a certain extent, and so do the current fluctuations induced in the "vacuum." These interactions cause the slight change of the magnetic moment. The numerical result is in excellent agreement with recent experimental measurements [2]. The magnetic moment of the electron was determined with great accuracy from the Zeeman effect of some fine structure doublets. Although the correction of the g factor cannot be understood in simple terms, it represents the most important result of quantum electrodynamics since it deals with one of the fundamental properties of the free electron—its magnetic moment.

The great success in these two instances of the quantum-electrodynamical concepts proves that the fundamental ideas must contain a great deal of truth. The main achievement of the recent development consisted in finding an unambiguous and relativistically invariant way of separating those effects of the interaction between light and electron which can be interpreted as additional mass and charge from the other effects which give rise to observable phenomena. The additional mass and charge are contained in the observed values of m and e and can never be observed independently. It must not be forgotten, however, that these magnitudes are still infinite in this theory. This constitutes a warning that the interaction of the electron with light quanta of very high energy is not yet understood. Somewhere at very high energies, the internal structure of the electron must play an essential role in a future theory in a way which is completely unknown. This structure appears at present in the form of the arbitrary length a which we have introduced as a radius of the electron in order to make the mass and the charge of the electron finite magnitudes.

The importance of the recent developments lies in the recognition of the following fact: For problems dealing with atomic energies only mass and charge of the electron are "structure dependent" (meaning dependent on the value of a and going to infinity if a is chosen zero), whereas all other effects, such as scattering cross sections, energy levels, magnetic moments, etc., can be calculated without making any assumption regarding the structure of the electron.

There is perhaps some significance in the fact that the theory of the electron cannot be brought into a completely satisfactory form without introducing some new elements into the theory at high energies. It cannot be a pure accident that the charge of the proton and of the meson is equal to the electronic charge, or that the classical electron radius r_0 is almost equal to the range of nuclear forces. There must be a connection between quantum electrodynamics and the future theory of mesons and of the nuclear forces, which at present exists only in very rudimentary form. The tie between these theories should be of importance for the electron only at energies of the order of the meson rest mass or higher. This would be high enough (>100 Mev) to leave unchanged the results of the theory for atomic energies. One may hope that the understanding of this tie will solve the problem of the electromagnetic mass and of the induced charge of the electron.

In discussing the classical electron theory, we remarked that a scattering experiment testing the limits of the classical theory would have revealed the existence of positrons, a phenomenon which was of fundamental significance for the further development of the theory. An experiment trying to test the present theory at high energies (100 Mev and over) will probably give rise to meson production. This is perhaps an indication of the important role of the mesons in a future theory of the electron. Future experiments with the new accelerating machines which are now under construction will reach energies of these critical values. It is hoped that the phenomena found by means of these new tools will shed new light upon the fundamental problem of the relation between elementary particles.

Remarks Added by the Author, 1972

Remark to p. 117. It is easy to understand what is going on when the vacuum is polarized around a protonic charge. Assume that the charge of the "bare" proton is e_0, which is assumed to be very much larger than its observed charge e, $e_0 \gg e$, and assume that the proton has a radius a very much smaller than the Compton wavelength λ_c of the electron, $a \ll \hbar/m_e c$. The charge e_0 (being positive) attracts the "vacuum electrons" which will gather as near as they possibly can around the proton. The total accumulated negative charge around the bare proton will be $-(e_0-e)$, so that the overall effective charge of the proton is

e. Each electron cannot be localized better than its wavelength λ. Hence a very large number of "vacuum electrons" whose wavelength is less than a will gather within the proton and cancel out most of the charge e_0 of the proton. The slower ones $\lambda > a$, however, come as near as they can, and they form a cloud of negative charge outside the proton. This explains that the induced charge around the proton is negative. A more accurate calculation shows that the fast electrons which are pulled into the sphere a have a total charge that is roughly $-ef(e^2/\hbar c) \log(\lambda_c/a)$; the negative cloud outside a reaches into distances λ_c but contains only a charge of the order $-e(e^2/\hbar c)$, say $g \cdot e^3/\hbar c$, where f and g are of the order one. Hence the bare charge e_0 must be

$$e_0 = e[1+(e^2/hc)(f \log(\lambda_c/a)+g)]$$

in order to make the total resulting observed charge of the proton equal to e. If one penetrates from outside toward the proton, the electric field of the proton seemingly increases when one gets nearer than λ_c; part of the negative charge remains outside and the charge inside must be correspondingly higher.

Remark to p. 122. Since this article appeared, the techniques of measuring the "Lamb shift" have vastly improved. Today it can be measured with an accuracy of 0.1 megacycle. The result of the measurements is equal to the calculated value to within the experimental error. Thus the effect of both the zero-point oscillations and the vacuum polarization have been verified. The vacuum polarization was also observed separately by measuring the energy of negative μ-mesons (heavy electrons) bound to atomic nuclei. A "muon" in a bound state of nonzero angular momentum is under the influence of the Coulomb field of the nuclear point charge (the finite extension of the nucleus can be neglected because the muon does not come into the immediate neighborhood of the nucleus in a state with $l \neq 0$). On the other hand the large mass of the muon causes the orbit to fall within the distance λ_c from the nucleus, just where the vacuum polarization caused by the nuclear charge is important. (The Lamb shift for a heavy electron with $l \neq 0$ is negligible.) Thus the energies of these muon states must show small deviations from the Balmer values. These deviations were observed and again have borne out quantitatively the existence of the vacuum polarization.

Remark to p. 123. I wish I could report that today—20 years later—there exists a qualitative *anschaulich* explanation of the fact that the magnetic moment of the electron is slightly larger than the value $e\hbar/2\,mc$ which follows from the Dirac equation without taking into account the effects of the vacuum. At first one might think that the magnetic moment should become smaller, since the interaction of the electron with the vacuum fluctuations would mix the opposite spin state into the state of the electron. This effect is certainly present [4] but seems to be overcompensated by another effect, which increases the magnetic moment.

Remark to p. 125. The expectations expressed at the end of this article did not turn out to be correct. Electrodynamics has been tested since then at energies up to 20,000 Mev and to distances as small as 10^{-15} cm. No deviation was found, and this indicates that the "radius" of the electron, if it exists at all, is smaller than that length and certainly much smaller than the Compton wavelength of the μ-meson. No connection between electrodynamics and nuclear phenomena has been established yet, and the similarity of the classical electron radius e^2/mc^2 with the range of nuclear forces is still unexplained.

References

[1] W. E. Lamb, Jr., and R. C. Retherford, *Phys. Rev.* **72,** 241, 1947.

[2] Nafe, Nelson, and Rabi, *Phys. Rev.* **71,** 914, 1947; D. Nagel, Julian, and J. Zacharias, *Phys. Rev.* **72,** 971, 1947; P. Kusch and H. M. Foley, *Phys. Rev.* **72,** 1256, 1947.

[3] This development started at a conference of theoretical physicists in June, 1947, on Shelter Island, New York, sponsored by the National Academy of Sciences. The same development has been carried out completely independently by a group of Japanese physicists around Professor Tomonaga. The following papers have been published so far: S. Tomonaga, *Prog. Theor. Phys.* **1,** 27, 1946; Koba, Tati, and Tomonaga, *Prog. Theor. Phys.* **2,** 101, 198, 1947; S. Kanesawa and S. Tomonaga, *Prog. Theor. Phys.* **3,** 1, 1948; S. Tomonaga, *Phys. Rev.* **74,** 224, 1948; H. A. Bethe, *Phys. Rev.* **72,** 339, 1947; H. W. Lewis, *Phys. Rev.* **73,** 173, 1948; H. A. Kramers, *Solvay Report,* 1948. J. Schwinger, *Phys. Rev.* **73,** 415, 1948; *Phys. Rev.* **74,** 1439, 1948; *Phys. Rev.* **75,** 651, 1949. R. P. Feynman, *Phys. Rev.* **74,** 939, 1430, 1948. F. J. Dyson, *Phys. Rev.* **73,** 617, 1948; *Phys. Rev.* **75,** 486, 1949. N. M. Kroll and W. Lamb, *Phys. Rev.* **75,** 388, 1949; T. Welton, *Phys. Rev.* **74,** 1157, 1948; J. B. French and V. F. Weisskopf, *Phys. Rev.* **75,** 1240, 1949.

[4] We are following here a calculation outlined by T. Welton, *Phys. Rev.* **74,** 1157, 1948.

[5] J. Mack, *Phys. Rev.* **73,** 1233, 1948.

[6] H. Kopfermann and W. Paul, *Naturwiss*, 1948.

The Compound Nucleus

1. Rarely has a single paper dominated a field of physics as has Bohr's address [1] to the Copenhagen Academy in 1936, in which he proposed the idea of a compound nucleus. During the 18 years since its appearance, it has been the decisive influence on the analysis of nuclear reactions.

What was the situation in nuclear physics when it appeared? Only a few qualitative facts were then known about nuclear reactions. Early work had shown that most cross sections were of the order of nuclear dimensions. (For charged particles the nuclear effects were even less prominent because of the Coulomb field.) More recently, Fermi and his collaborators [2] had discovered much larger cross sections for slow neutron reactions in some elements.

Previous theoretical attempts [3] to explain this variation in the size of cross sections were all based on an extremely simple picture of the nucleus, the potential well model. According to this model the effect of the target nucleus upon an incident particle can be described, at least as a first approximation, by an attractive potential. The quantal state is given by

$$\psi = \phi(r)\chi(r_1 \ldots r_A) \tag{1}$$

where χ is the wave function of the target nucleus and $\psi(r)$ is the wave function of the incident particle. It is assumed that χ is uninfluenced by the interaction, while $\phi(r)$ is supposed to be the solution of a one-particle problem in which the particle moves in the potential $V(r)$. In the simplest potential well model $V(r)$ is the square well:

$$\left.\begin{array}{ll} V = -V_o & r < R \\ V = 0 & r > R \end{array}\right\} \text{with } R = r_o A^{1/3}. \tag{2}$$

As usual, r_o is a constant of the order of 10^{-13} cm and A is the

With F. L. Friedman. Revision by the author of an article originally in *Niels Bohr and the Development of Physics*, W. Pauli et al., eds., Pergamon Press, London, 1955.

mass number of the target nucleus. V_o is of the order of nuclear energies, a few tens of Mev.

The predictions of the potential well model are also simple. Only two types of reaction can be exhibited: elastic scattering and radiative phenomena. The scattering cross section dominates the scene. It is generally of nuclear dimensions, but it attains large values at widely spaced resonances. The spacing for a given angular momentum l is about 10 or 20 Mev, and at resonance the scattering cross section is approximately $(2l+1)$ $4\pi \lambdabar^2$. If a resonance occurs at low neutron energies, the scattering cross section attains very large values.

At first the large resonance cross sections seemed to be an explanation of the very high cross sections which Fermi and his collaborators found with thermal neutrons in some nuclei. However, there are other consequences of the potential well model which soon disqualified it: (1) The capture cross section is generally very small on this model and even at resonance the scattering dominates. In this model the neutron spends only a short time within the potential, and hence the probability of a radiative transition to a bound state within the potential is extremely small. (2) The wide spacing of the resonances reflects the most characteristic feature of both cross sections: They change slowly with energy. When $\lambdabar^2$ is divided out, they change but little within energy intervals of one Mev or less. This slow change with energy and the predominance of scattering even in resonance are typical of single particle behavior.

Nature disagrees with the potential well model on both counts. Just before Bohr's celebrated address, experiments performed by Bjerge and Westcott, Moon and Tillman, Szilard, Fermi, and others [4] showed that neutron cross sections vary greatly within a few electron volts: Resonances are both extremely narrow and closely spaced. At the same time the cross sections at resonance turned out to be mainly capture.

In 1936, then, the potential well model was due for replacement. The narrow close-spaced resonances and the high capture probability in resonance required some changes in the description of nuclear reactions. In fact, a complete change of view resulted, and it was this change that was the subject of Bohr's address.

The main concept Bohr introduced is the compound nucleus: A many-body state formed immediately after the impinging particle hits the nucleus, it is the antithesis of the single particle model it replaced. The argument runs as follows: The observation of closely-spaced resonances in heavy nuclei is an indication that the states formed in the reaction are states of a many-particle system. The form (1) cannot describe reactions where closely-spaced resonances occur. Level distances of a few volts in systems of nuclear size can only occur if a large number of particles are involved in the excitation. It is wrong therefore to assume that the passage of the incident particle through the nucleus does not disturb the state of the target nucleus appreciably. On the contrary, in order to explain the excitation of many particles, it is natural to go to the opposite extreme. The assumption is therefore made that all nucleons forming the nucleus or incident upon it interact strongly.

As a consequence of the strong interaction, it was assumed, the incident particle and the nucleus it strikes coalesce, forming a compound state in which all or very many nucleons participate collectively; the resonances are the energy values of the quantum states of the compound system. These states are not strictly stationary; they have a finite lifetime, since they can decay by the re-emission of the incident particle, by γ radiation, or otherwise. The width of the resonances indicates, however, that the lifetimes of the compound states created by particles of low energy are extremely long compared to the straight passage time of the incident particle across the nucleus. The states are almost stationary and should not differ greatly in their general properties from the real stationary states of the compound system at somewhat lower energy.

This new view was extremely successful in describing the low-energy neutron experiments. Not only are closely spaced and narrow resonances explained, but also the actual predominance of capture over scattering in low energy resonances is easily described. The very long lifetime of the compound state makes it possible for electromagnetic radiation to compete successfully with other modes of decay. The major part of the resonance cross section is thus shifted from scattering in the potential well model to capture on the Bohr picture.

At higher energies the resonances increase in width and start to overlap. The observations are easily explained by two facts: First, the emission probability of a particle increases with energy, and, second, more reaction channels are open at higher energy, each one contributing to the total width. Present experimental material and reasonable extrapolations indicate that sharp and well-defined resonances are found only up to an energy of the incident particle of several Mev except for very light nuclei. At higher energies the widths become comparable to or larger than the level distance, and the resonance structure is lost.

2. At the same time as Bohr gave his general description of nuclear reactions, more quantitative treatments of resonance phenomena were initiated by Breit and Wigner [5]. Many generalizations of this work have been carried out, the most important ones by Bethe and Placzek [6] and later on by Wigner and his collaborators [7]. Accurate measurements during the last two decades have shown that the resonance phenomena in nuclear reactions are well represented in terms of the Breit-Wigner formula. The accuracy of this representation is a proof of the existence of well-defined compound states, and the prop-

erties of these states—lifetime, decay probabilities into different channels, etc.—can be measured and systematized.

In the region of overlapping compound levels, the Breit-Wigner formula must be extended to include many levels, and the description becomes impractical, since the resulting cross sections depend largely upon unknown phase relations between the resonances. In order to allow some practical conclusions to be drawn about the yield of nuclear reactions in that region, the picture must be simplified.

This simplification is usually introduced by an assumption which Bohr suggested in his address. We divide the nuclear reaction into two steps, the formation of the compound nucleus and its subsequent decay. Then we make the assumption that the decay is independent of the mode of formation of the compound nucleus. According to this "independence hypothesis" it should not matter what incident particle and what target nucleus are used as long as the same compound system is formed.

In the resonance region, the Breit-Wigner formula factors automatically into the two stages of Bohr's description, the cross section for the formation of the compound nucleus and the probability of its decay into a given final configuration. In the region of overlapping resonances the factorization is a special assumption. With this assumption the cross section of an (a, b) reaction can be written in the form

$$\sigma(a,b) = \sigma_c(a) \, . \, (\Gamma_b/\Gamma)_c \tag{3}$$

where $\sigma_c(a)$ is the cross section for the formation of a compound nucleus by the particle a and $(\Gamma_b/\Gamma)_c$ is the probability that the particle b is emitted by the compound state.

This formula can be used for calculating reaction cross sections. The cross section $\sigma_c(a)$ can be roughly determined by assuming that the compound nucleus is formed immediately when the nuclear surface is reached. All one has to do is to compute the probability for the incident particle to reach the surface of the nucleus, a problem which can be solved by simple quantum-mechanical calculation [8]. Next the decay probability of the compound nucleus can be determined by reversing the process. Because the decay through a given channel is the inverse of the formation of a compound nucleus through the same channel, any method that allows the calculation of $\sigma_c(a)$ can also be used to compute the factor $(\Gamma_b/\Gamma)_c$ in [3].

The method of calculating reaction cross sections just sketched is often referred to as the "statistical method." Some of its conclusions can be suitably expressed by means of thermodynamical concepts which are based upon Bohr's picture of the sharing of energy between the constituents of the compound system. The excited compound nucleus is considered as a heated system and its subsequent decay as evaporation of particles [9].

The statistical method of determining the yield of nuclear reactions gives a reasonable account of their most important features. For example, it follows that reactions initiated by protons are weaker than those initiated by neutrons by a factor corresponding to the barrier penetration. A similar factor is expected to appear in the ratio between the yields of two reactions initiated by the same particle, one leading to proton emission and the other to neutron emission. Furthermore, the average energy of the emitted particles is expected to be small compared to the total energy available, the balance being left as excitation of the residual nucleus.

These predictions are qualitatively correct. Reaction cross sections as functions of the energy of the incident particle, in particular, the yield-energy curves of proton or α-initiated reactions, are reasonably well explained [10]. The relative yields of (x, n) and $(x, 2n)$ reactions (x being neutron, proton, or α-particle) are reproduced in their essential features. Also the energy distribution of reaction products is predicted fairly well. The spectrum of neutrons and protons emitted from nuclei bombarded with neutrons of 14 Mev or with protons of similar energy fits approximately the predicted Maxwell distribution of an evaporating compound nucleus, and the "temperatures" are not too far from the expected magnitude [11].

The success of the statistical method is limited to the qualitative description of these salient features. The accumulation of more quantitative data in recent years has produced a growing number of quantitative exceptions, and certain phenomena exhibit gross features unparalleled in the statistical model. Disagreements have been found, for example, in the study of the energy dependence of total neutron cross sections and in the yield of those reactions in which charged particles are emitted. When they are observed with poor energy resolution, the neutron cross sections show an energy dependence which is vaguely similar to the energy dependence of the scattering at a potential well [12], a behavior which is unexpected on the basis of the statistical model; charged particles frequently are emitted with energies much larger than the ones expected from evaporation and with an angular distribution peaked in the forward direction [13].

These disagreements and others, some of which will be discussed later on, are serious enough to warrant a thorough analysis of the two assumptions on which the statistical treatment of nuclear reactions was based: the independence of the decay of the compound nucleus from the way in which it is formed; and the immediate formation of the compound nucleus when the incident particle reaches the nuclear surface.

3. In examining the assumptions underlying our picture of nuclear reactions, we begin with the independence hypothesis. After surveying the experimental evidence that can be brought to bear on this assumption, we shall have a look at its logical foundations.

There is not very much experimental material available which can be used to test this assumption directly. It is not easy to reach the same energy region in the compound system with different reactions. Nevertheless a few statements can be made.

As long as the energy region reached in the compound nucleus is in the resonance region, the independence hypothesis has always proved correct [14]. At higher energies the results are more questionable. Ghoshal [15] has shown that the independence assumption seems to hold for a compound nucleus of an excitation energy of 15 to 40 Mev when produced by protons on Cu^{63} or by α-particles on Ni^{60}. In many other cases, however, there are indications that the assumption is not very good. B. Cohen and his collaborators [16] have pointed out that the decay probabilities of the compound nucleus sometimes depend quite critically upon the way the compound state is formed. For example, Cohen and Newman [16] compare the relative probabilities of emission of protons and of neutrons from compound states formed either with protons or with neutrons. Nuclei with mass numbers in the region 48 to 71 were bombarded with protons of 21 Mev and neutrons of 14 Mev. When the reaction is initiated by protons, it turns out that proton emission is more probable than neutron emission.

Furthermore, the independence assumption is indirectly attacked by other evidence. In Bohr's picture, independence is related to the idea that the energy is shared by all particles and that the direction of incidence or position in the nucleus of the particle originally initiating the reaction becomes irrelevant in the compound state. In fact, as was mentioned, one often finds reaction products emitted with a much higher energy than one would expect if the energy of the incoming particle were shared among all constituents; and those reaction products often show an angular distribution (mostly forward) with respect to the direction of the incident particle. In these cases the compound state has more memory of the initial process than seems generally compatible with the independence hypothesis.

What logical justification is there for the independence assumption, and what are the reasons for its possible breakdown? If the energy of the incident particle is within the region of sharp and well-defined resonances and if the energy coincides with or is near to a resonance, the assumption is obviously justified. The nuclear reaction then produces only one quantum state of the compound system. The properties of a given quantum state evidently are independent of the way it is produced.[1] The validity of this conclusion is limited only by the fact that the resonances are not stationary states in the strict sense because of their finite width. Actually, because of the overlapping of the wings of neighboring resonances, it is never exactly true that only one quantum state is realized. However, the deviations are small and of the order of level width to level distance.

In the region of considerable overlap of resonances the validity of the independence assumption is by no means obvious.[2] In this case an incident particle of fixed energy excites several compound states, and the relative phase of the states will depend upon the nature of the excitation. Hence, if the same compound nucleus is excited to the same energy by different processes, one would expect different phase relations between the compound states and different modes of decay, since the emission probabilities of a linear combination of states depend upon their phase relations. Thus in the region of overlapping compound states the independence assumption is no longer trivial; on the contrary, it is in need of justification (of a different justification from that of an isolated state). Otherwise, it is invalid.

When the density of overlapping states is very large and many compound states are excited simultaneously, a new situation

[1]Such a state may show directional "memory," but not forward asymmetry. It has a symmetry plane at right angles to the incident direction. (Polarization of the original beams or targets is assumed absent.)
[2]Substantially, the following arguments were presented in 1939 by Bohr, Peierls, and Placzek [17].

may arise. So many states and, therefore, so many phases are involved in one reaction that, even though the relative phases are determined by the excitation process, they may act in respect to the decay process as if they were random. It is then possible that the second stage of the nuclear reaction appears independent of the first in the region of strong overlap.

That this possible independence should indeed be realized was made plausible very early in the form of a quasi-classical argument to which we have already alluded. It runs as follows: The energies for which overlap of resonances occurs are high enough that classical considerations may be significant. Using classical considerations alone, we argue first that the incident particle and the constituents of the target nucleus interact so strongly with each other that, in a time short compared to the free passage time of the particle through the nucleus, the energy is distributed among a large number of constituents. Second, since the particles are constantly and rapidly exchanging energy, a state of statistical equilibrium is reached before the compound state is broken up; and finally, because decay takes place from a statistical distribution, the probabilities of the different break ups do not depend upon the nature of the initial delivery of the energy.

The quasi-classical picture also contains an explanation of how exceptions to the independence assumption arise. The picture is only valid if the energy of the incident particle is high enough so that classical reasoning may be employed. On the other hand, because high incident energy gives rise to relatively short-lived compound states, it is not certain that there exists any energy region for which classical reasoning is applicable and in which the lifetime is long enough so that statistical equilibrium is reached before break up.

Not only does the time available for reaching equilibrium fall with increasing energy, but the interaction cross section between the incident particle and the nucleons becomes smaller, hampering the establishment of thermal equilibrium. Within nuclear matter the mean free path of nucleons of 100 Mev or over becomes comparable with the nuclear radius, and some nuclear reactions at that energy have been successfully described by assuming that an energetic nucleon passing through the nucleus interacts with *individual* nucleons rather than with a many-body system [18]. In such a reaction the details of "formation" are all important.

At intermediate energies where the interaction is still strong, the extreme description of nuclear reactions as individual collisions, nucleon by nucleon, will not hold. Nevertheless, it is not necessary to return to the opposite case of states weighting all nucleons equally. Something in between is reasonable, and one can easily imagine mechanisms that might lead to nuclear reactions in which the break up of the compound system is dependent on the mode of formation.

For this intermediate energy region, Bethe [19] suggested the possibility of a "spot heating" effect. When the incident particle impinges upon the nuclear surface, its energy is first shared by the nucleons surrounding the point of contact.[3] This small region of the nucleus acquires a rather high temperature, and it may emit a nucleon immediately with an energy much higher than that expected if the total energy were distributed over the whole nucleus. Somewhat different from but related to spot heating is the transfer of energy directly to a single nucleon when an incident particle grazes the surface. In both cases the energy of the ejected particle will be large, but spot heating corresponds crudely to the backward particle emission and grazing collisions to the forward part of the angular distribution. Another possible mechanism of this type would be the setting up by the initial

[3] All localizations in these quasi-classical discussions are defined only with an accuracy of one wavelength, which is small compared to nuclear dimensions for energies of 10 Mev and over.

impact of a surface deformation which then travels around the nucleus and is focused on the opposite side where it might then give rise to the detachment of a particle. The reactions produced by these mechanisms are all examples in which the independence assumption breaks down.

In his original paper Bohr carefully pointed out that the compound nucleus and the two-step analysis of reactions would not provide an adequate picture for reactions involving very light nuclei. Deuteron stripping and pickup reactions appropriately fall in this class even though the idea of a compound nucleus involving one of the nucleons transferred in the reaction is a useful concept in the detailed theory of these reactions. As the compound system often does not form in these reactions, they play a multiple role in this classification, falling alongside spot heating and grazing in their failure to share the incident energy statistically.

It is interesting to reanalyze the independence hypothesis in the following terms: The quasi-classical and the quantum description of the compound system are related by the correspondence principle. In particular, the average energy spacing D of the quantum states in systems of simple periodic character corresponds to the period $\tau \cong 2\pi\hbar/D$ of the motion. In complicated systems, one cannot easily define the period of the motion; however, the time $\tau = 2\pi\hbar/D$ still plays the role of the time interval in which the classical motion has gone through all those configurations which are compatible with the initial conditions: After a time of the order τ the configurations will all be similar (within the uncertainty relations) to configurations which were realized before [20].

In the region of strongly overlapping compound states, the width Γ of the states is much larger than the distance D, which means that the lifetime $\tau_c = \hbar/\Gamma$ is much shorter than the characteristic repetition time τ. In other words, the system has not enough time to go through all configurations compatible with its initial energy, angular momentum, etc. It would go through all these configurations if, by some device, all channels of decay were closed. Further, if the channels leaked so slowly that all configurations were passed through many times, the decay of the compound system would be independent of the exact starting point. Both the decay time and decay probabilities would be independent of formation. With these channels open, however, the lifetime becomes so short that only a few of the configurations can be realized, and when the system passes through few configurations, just which ones are realized may depend critically upon the initial conditions. Therefore the decay of the compound nucleus may depend upon the way it was formed.

Again, we find that, if the independence hypothesis is valid in the region of overlapping, it must be true despite the fact that different configurations are realized for different modes of formation. Since the probabilities of decay are general features, it is possible that they may not depend on the detailed configurations; they may come from averages which wipe out some of the details. Nevertheless, above the energy region of sharp separated resonances, the independence hypothesis is insecure.

In general, then, the independence hypothesis is a very far-reaching assumption. Its significance is different in the region of sharp resonances and in the region of overlapping widths. In the resonance region the hypothesis is obviously valid, since it follows from the fact that well-defined quantum states are created as indicated by the sharp resonances. At higher energies, however, its validity is doubtful. There is a middle region where the widths of the resonances are of the order of their spacing in which the independence hypothesis will not hold at all. There only a few component states are simultaneously excited and the phase relations necessarily will be typical of the mode of excitation. Finally, in the region of strong overlapping the independence assumption might be valid again, at least in some cases.

In these cases its validity represents a limiting case of statistical disorder reached in the compound state before its break up; and, since decay times decrease rapidly with increasing energy, it is not certain whether these chaotic conditions ever are reached.

It is, therefore, understandable that nuclear reactions have been found for which the two-step description with independence of decay from formation of the compound system is not applicable. Such nuclear reactions in which the properties and probabilities of the products are closely related to the details of the initial configuration are to be expected. Fruitful idea though it was, independence is too drastic a simplification.

4. We now turn to the second hypothesis used in the statistical method of calculating nuclear reactions, the assumption that a compound nucleus is created when the incident particle reaches the nuclear surface. There cannot be any doubt that compound states are sometimes created by the incident particles; the observation of closely spaced resonances is the best proof. The remaining question is whether such many-particle states are always formed "immediately" after the particle enters the nucleus.

The assumption of immediate formation was based upon the classical picture of a particle impinging upon a system of strongly bound constituents. The strong interaction energy between nucleons seems to lead necessarily to a quick exchange of energy. The recent success of the shell model in describing the properties of the lowest states of the nuclei casts some doubt upon this conclusion. The evidence revealed in the light of the shell model shows that a nucleon can move comparatively freely within the nuclear volume. Such a nucleon seems to possess an angular momentum of its own and to move in a well-defined single particle orbit.

As yet no satisfactory explanation of this behavior exists. The apparent contradiction between single particle orbits and the strong interaction observed in the scattering of nucleons by nucleons might be explained in two ways: Either the forces between nucleons are much weaker when they are within nuclear matter, or the configurations in low nuclear states are such that they are describable by independent particle orbits in spite of the strong interactions.

Not very much evidence has been found so far in support of the first explanation. Indirectly, perhaps, our present difficulties in understanding the saturation of nuclear forces on the basis of the free nucleon-nucleon interaction speak in favor of some change in the internucleon potential when the nucleons are closely packed. However, it seems that small changes would be sufficient to explain saturation, for example, the introduction of repulsive three-body forces, or a large repulsive core in the potential between two particles [21]. These changes are not in themselves sufficient to guarantee the free particle behavior of nucleons in nuclear matter, since they do not exclude strong interactions between two neighboring nucleons. More drastic changes have been suggested, which would make the free particle motion obvious. An example is a nonlinear behavior of nuclear potentials. A saturation value of the potential is assumed to be created by a high density of particles [22]. If this saturation value is reached within nuclear matter of normal density, the nucleons fail to interact, even if they approach each other closely.

The second explanation has not yet been formulated in a satisfactory form. The free particle behavior in low excited states might perhaps be understood even in the presence of strong interaction by invoking the exclusion principle, which forbids momentum or energy transfer from one particle to the other because the states into which the particles could be scattered are all occupied [23]. However, until it is proved that the free particle states are indeed the lowest, this reasoning is not much more than a plausibility argument.

In spite of the absence of a satisfactory explanation, there is no doubt that the low-lying nuclear energy states can be described surprisingly well by a system in which nucleons move with little interaction in a common potential well; and the many facts thus summarized require a reappraisal of our idea of the formation of a compound nucleus by an incident particle impinging on a target nucleus.

In attempting this reappraisal, we may be guided by some of the discrepancies between the experiments and the statistical model. For example, as we mentioned, the large-scale energy dependence of the total neutron cross sections disagrees with the statistical theory. The statistical model for neutron cross sections assumes that the compound nucleus is formed immediately when the incident neutron reaches the nuclear surface; and the cross section for reaching the surface turns out to be a monotonically decreasing function of the energy, going as $E^{-1/2}$ for small energies and reaching the asymptotic value $2\pi R^2$ for large energies [8]. The experiments do not agree; the observed cross sections show a more complicated behavior which seems to point to some combination of single particle and compound nucleus. Looked at under the microscope of high resolution experiments, the cross sections show the narrow peaks which led to the Bohr theory. Looking more broadly, Barschall and his coworkers [12] have plotted the observed total cross section in a three-dimensional plot against energy E and atomic number A. This plot exhibits systematic regularities with maxima and minima at values of E and A where the old pre-Bohr theories of the simple potential well would have placed them.

Although these large-scale maxima are not as strongly pronounced as the simple well theory says, they are entirely unexpected on the statistical model. They strongly suggest a partial return to the old potential well. Obviously, since slow neutron experiments show that compound states are formed in which the unit length of path for the incident particle in nuclear matter to form the compound system. For an incident particle of energy E, the coalescence coefficient is given by

$$K = \left[\frac{m}{2(E+V_o)}\right]^{1/2} 2(V_1/\hbar). \tag{4}$$

$(K)^{-1}$ represents the distance which the particle must traverse within the nucleus before the compound formation occurs with appreciable probability, and $\frac{1}{2}(\hbar/V_1)$ is the mean time before coalescence takes place. On this picture, if an incident particle gets into the nucleus, it is reflected back and forth approximately as in the old potential well model before escaping or coalescing. Where the old well has virtual states, they have been reintroduced as a precursor to the final compound nucleus; and the nuclear reaction may be thought of as proceeding in two stages, a brief formation of a single particle state followed by escape or coalescence.

In discussion some years ago when the evidence for the shell structure was accumulating and some of the inadequacies of the compound nucleus picture were becoming more apparent, Bohr suggested that these new insights would give us a more detailed picture of nuclear reactions, and, in particular, that one should investigate the possibility of introducing earlier stages into the picture of the reaction before the formation of the eventual compound state. The intermediate stage introduced by the optical model into the picture of a nuclear reaction, the stage during which the single particle is inside the nucleus, is probably the simplest realization of the stages suggested by Bohr. During this stage the optical model automatically combines the reflections of the waves at the edge of the potential well with the coalescence inside the nucleus to tell us how often the compound system is formed.

energy is distributed over the nuclear constituents
body description cannot be exactly valid; but an i
situation between the pure one-body treatment a
mediate formation of the compound nucleus may be
In this intermediate theory, for some circumstances a
motion of the incident particle within the nucleus s
approximately the same as a single particle moving in a
well.

5. One attempt to combine single particle and com
nucleus pictures is embodied in the optical model of the n
[24]. This model describes the effect of the nucleus o
incident particle by a potential well, $-V_o(r)$, but allows fo
possibility of formation of the compound nucleus by addin
the potential a negative imaginary part, $-iV_1(r)$. This p
produces an absorption of the incoming wave within the nucle
and this absorption is supposed to represent the formation of
compound nucleus. As V_1 measures what is taken out of th
single particle description rather than what goes into any other
particular mode, our use of the term "formation of the compound nucleus" is broad here. It includes not only processes in which the particle shares its energy with all nucleons and forms a compound state in the orthodox sense, but also processes of the kind described in Section 3 in which the particle interacts with parts of the target only. The essential point is that it represents any process which removes the incoming particle from the entrance channel.

On the optical model, compound formation does not take place "immediately" nor with complete certainty. Even if the incident particle has entered the nucleus, it is removed from its free particle state only with some delay and a certain probability. If $V_o(r)$ and $V_1(r)$ are reasonably constant over the nucleus, one can define a coalescence coefficient; it is the probability per

With suitable simple assumptions regarding the form of V_o and V_1 the effect of the potential

$$V = -[V_o(r) + iV_1(r)]$$

on an incident beam of particles can be determined. The scattering and absorption cross sections, σ_{el}^{op} and σ_a^{op}, are thus obtained as functions of the energy. The absorption cross section σ_a^{op} of the model should represent the cross section for the formation of the compound nucleus in real life. The scattering cross section σ_{el}^{op} derived from the model must be related to the elastic scattering. By comparing the cross sections calculated from the model with experiments one hopes to be able to determine $V(r)$.

When the observed total neutron cross section is averaged over an energy interval large enough to smooth out any narrow resonances, it is surprisingly well represented by the complex potential model. With the simplest assumption of a square well for both V_o and V^1, one can reproduce the characteristic maxima and minima of this average cross section as a function of energy and mass number, covering all A and an energy range from zero to a few Mev. The following potential gives the best fit [25] (see the Additional Remarks at the end of this paper):

$$\left.\begin{array}{ll} V_o = 40 \text{ Mev} & \\ & \text{for } r < R \\ 1 \text{ Mev} < V_1 < 2 \text{ Mev} & \\ \text{and} & \\ V_o = V_1 = 0 & \text{for } r > R \\ \text{with} & \\ R = 1{\cdot}45 \times 10^{-13} A^{1/3}. & \end{array}\right\} \quad (5)$$

We conclude therefore that the mean free path for compound nucleus formation of a slow neutron entering the nucleus is about

$1-2\times10^{-12}$ cm.

The neutron does not form a compound nucleus immediately upon entering the target nucleus; in fact, the probability of formation is about 0.30 for one trip across a medium-sized nucleus.

6. That such a simple model as the complex potential fits the experimental data so well is perhaps a bit surprising. In order to see whether or not the correspondence can be understood, it is necessary to examine the relation of reality to the optical model more closely than was done above. In particular, just what the formation of the compound nucleus means needs more careful consideration.

In the optical model, compound nucleus formation and absorption are the same and represent the removal of the particle from the entrance channel. Reality is certainly more complicated. Among other complications, the compound nucleus can emit the incident particle back into the entrance channel.

This re-emission does not lead to any trouble in the high-energy region. If the incident particle arrives in the continuum rather than the resonance region, so many channels are open for decay of the compound system that the probability of re-emission is negligible. Compound nucleus formation and absorption can therefore be equated. In this respect "reality" and the optical model are the same.

On the other hand, if the energy of the incident particle is low, the reality of closely spaced resonances does not correspond with the smooth energy dependence of the optical model, and the existence of resonances must be expressed by boundary conditions on the incident wave which are violently different from those of the optical model. (For example, the conditions for resonance require zero slope of the external wave function at the nuclear radius [26].) Also fewer decay channels are open, and we can no longer be sure that re-emission into the entrance channel is negligible. Consequently, compound nucleus formation can no longer be equated with absorption.

In the resonance region, then, there are conceptual difficulties in relating the optical model to the actual state of affairs. In order to resolve these difficulties, it is necessary to carry out some kind of averaging over the resonances. Such averaging is required to wash out the sharp bumps. It also can be used to overcome the difficulty in relating compound nucleus formation to absorption when there is re-emission into the entrance channel. Finally it will supply us with a resolution of the paradox represented by the difference between resonance boundary conditions changing rapidly with energy and the slow energy variations of the optical model.

How the averaging over resonances arrives at a consistent relation between "reality" and the optical model can be seen in two ways—one somewhat mathematical, the other relatively *anschaulich*. Because the more mathematical method allows us to make a reasonably concise statement of the relation, we shall treat it first and then try to see into the results through the more intuitive discussion.

For simplicity, we restrict the discussion to neutron reactions with the $l = 0$ partial wave. A wave function

$$\psi = \frac{A}{r}(e^{-ikr} - \eta e^{ikr}) \tag{6}$$

outside the range R of interaction with the nucleus is then sufficiently general if η is a function of the incident energy E, a function which varies rapidly as E passes through a resonance. The cross sections are related to η as follows [26]:

$$\sigma_{el} = \frac{\pi}{k^2}|1-\eta|^2,$$

$$\sigma_r = \frac{\pi}{k^2}(1-|\eta|^2),$$

$$\tag{7}$$

and

$$\sigma_{\text{tot}} = \frac{\pi}{k^2}2(1-\mathscr{R}(\eta)),$$

where $\mathscr{R}(\eta)$ means the real part of η. These are the genuine elastic scattering, reaction, and total cross sections for that precise energy E of the incident neutrons which corresponds to the wave number k: i.e., $E = (\hbar k)^2/2m$.

We now average these cross sections over an energy interval I which contains many resonances. The average is defined for convenience so that $k^2\sigma$ is averaged rather than σ, but when the mean energy is great compared with the interval I, the k^2 can be factored out; only at very low energies is some special consideration needed. With this caveat, we define

$$\langle f(e)\rangle = \frac{1}{I}\int_I f(\varepsilon)\, d\varepsilon$$

and

$$\bar{\sigma} = \frac{1}{k^2(E)}\langle k^2\sigma\rangle.$$

From these definitions we obtain

$$\bar{\sigma}_r = \frac{\pi}{k^2}(1-\langle|\eta|^2\rangle) = \frac{\pi}{k^2}(1-|\langle\eta\rangle|^2)-\sigma_{fl} \tag{8}$$

where

$$\sigma_{fl} = \frac{\pi}{k^2}\langle|\Delta\eta|^2\rangle$$

and $\langle|\Delta\eta|^2\rangle = \langle|\eta|^2\rangle - |\langle\eta\rangle|^2$ is the mean square fluctuation in the interval I of the coefficient of the outgoing wave. Also

$$\bar{\sigma}_{\text{tot}} = \frac{\pi}{k^2}2(1-\mathscr{R}(\langle\eta\rangle)),$$

which depends only on $\langle\eta\rangle$ as it must because $\bar{\sigma}_{\text{tot}}$ is linear in η. ($\bar{\sigma}_r$, however, depends on $\langle|\eta|^2\rangle$.) We call σ_{fl} the fluctuation cross section.

In making the correspondence between reality and the optical model, we must identify the total cross section found from the model, $\sigma_{\text{tot}}^{\text{op}}$, with the average of the "real" total cross section at precise energies. Part of our correspondence is therefore

$$\sigma_{\text{tot}}^{\text{op}} \equiv \bar{\sigma}_{\text{tot}} = \frac{\pi}{k^2}2(1-\mathscr{R}(\langle\eta\rangle)). \tag{9}$$

To complete the correspondence, we wish to identify the absorption cross section of the model with formation of the compound system. The absorption on the model corresponds therefore to the sum of the reactions and that part of the elastic scattering which occurs by decay of the compound system through thc entrance channel. The analogous cross sections in a resonance theory are

$$\sigma_c = \sigma_r + \sigma_{ce}$$

where σ_c is the cross section for compound formation, and (on the right) it is expressed as the decay into reaction cross section σ_r and compound elastic scattering $\bar{\sigma}_{ce}$. Performing the averages, we take

$$\sigma_a^{\mathrm{op}} \doteq \bar{\sigma}_c = \bar{\sigma}_r + \bar{\sigma}_{ce}. \tag{10}$$

We now proceed to the determination of σ_c. Using (8) and (10), we find

$$\bar{\sigma}_c = \frac{\pi}{k^2}(1 - |\langle\eta\rangle|^2) - \sigma_{fl} + \bar{\sigma}_{ce}. \tag{11}$$

At this stage we have no specific representation for $\bar{\sigma}_{ce}$, but we have the following information about it: At high energies $\bar{\sigma}_{ce}$ vanishes because of the competition with many modes of decay of the compound system. At the same energies we expect η to be smooth, and therefore the fluctuation σ_{fl} also becomes negligible. In this energy region therefore

$$\bar{\sigma}_c \approx \frac{\pi}{k^2}(1 - |\langle\eta\rangle|^2). \tag{12}$$

At lower energies both σ_{fl} and $\bar{\sigma}_{ce}$ increase. We shall see that (12) remains valid and that the two cross sections σ_{fl} and $\bar{\sigma}_{ce}$ are equal. To examine the validity of (12), in the low-energy region, we go to the extreme case of well-separated resonances ($\Gamma \ll D$).

We introduce the necessary extra information—a resonance theory—in the form of an approximate η good at an isolated resonance occurring in the compound system for the incident neutron energy E_s:

$$\eta_{BW} = e^{i2\delta}\left(1 - \frac{i\Gamma_n}{E - E_s + i\Gamma/2}\right) \tag{13}$$

where δ is a slowly varying phase changing only over the single particle energy spacing and therefore irrelevant as long as we stay within a reasonable energy range. For this special case, as we see by putting η_{BW} into (7), the reaction cross section is

$$\sigma_r = \frac{\pi}{k^2}\frac{\Gamma_n(\Gamma - \Gamma_n)}{(E - E_s)^2 + (\Gamma/2)^2}$$

in agreement with the Breit-Wigner formulae

$$\sigma_c = \frac{\pi}{k^2}\frac{\Gamma_n\Gamma}{(E - E_s)^2 + (\Gamma/2)^2},$$

$$\sigma_r = \sigma_c\frac{\Gamma - \Gamma_n}{\Gamma}, \quad \text{and} \quad \sigma_{ce} = \sigma_c\frac{\Gamma_n}{\Gamma}.$$

Upon averaging, we then obtain

$$\bar{\sigma}_c = \frac{\pi}{k^2}\frac{2\pi\bar{\Gamma}_n}{D}$$

where $\frac{\bar{\Gamma}_n}{D} = \frac{1}{I}\sum_s \Gamma_n^{(s)}$ is the average neutron width divided by the average spacing; and

$$\bar{\sigma}_{ce} = \frac{\pi}{k^2}\frac{2\pi}{I}\sum(\Gamma_n^{(s)})^2/\Gamma^{(s)}. \tag{14}$$

We compare the result for $\bar{\sigma}_c$ with the value of $\frac{\pi}{k^2}(1-|\langle\eta\rangle|^2)$ by putting in η_{BW} for η. We get

$$\langle\eta_{BW}\rangle = e^{i\delta 2}\left(1-\pi\frac{\bar{\Gamma}_n}{D}\right),$$

and therefore

$$\frac{\pi}{k^2}(1-|\langle\eta_{BW}\rangle|^2) = \frac{\pi}{k^2}\frac{2\pi\bar{\Gamma}_n}{D}\left(1-\frac{\pi}{2}\frac{\bar{\Gamma}_n}{D}\right).$$

Because $\Gamma_n \leq \Gamma \ll D$, the last term in the brackets is negligible. Consequently, Eq. (12) is valid even in this extreme case. (What is neglected in making the equality is of the same order as the accuracy of the approximate resonance theory we have used.) This result is equivalent (according to (11)) to

$$\sigma_{fl} = \bar{\sigma}_{ce},$$

an equality which also can be established directly by evaluating $\langle|\Delta|^2\rangle$ to determine σ_{fl} and then comparing the result with $\bar{\sigma}_{ce}$ as given by (14).

We are now in possession of the complete correspondence

$$\sigma_{\text{tot}}^{\text{op}} \doteq \frac{\pi}{k^2}2(1-\mathscr{R}(\langle\eta\rangle))$$

$$\sigma_a^{\text{op}} \doteq \frac{\pi}{k^2}(1-|\langle\eta\rangle|^2)$$

and consequently

$$\sigma_{el}^{op} = \frac{\pi}{k^2} |1 - \langle \eta \rangle|^2.$$

The whole answer can be abbreviated:

$$\eta^{op} \doteq \langle \eta \rangle.$$

A little more light may be shed on the identification of the η^{op} with $\langle \eta \rangle$, by considering the scattering averages,

$$\bar{\sigma}_{el} = \frac{\pi}{k^2} \langle |1 - \eta|^2 \rangle$$

$$= \frac{\pi}{k^2} |1 - \langle \eta \rangle|^2 + \frac{\pi}{k^2} \langle |\Delta \eta|^2 \rangle.$$

Therefore

$$\sigma_{se} \equiv \bar{\sigma}_{el} - \sigma_{fl} = \frac{\pi}{k^2} |1 - \langle \eta \rangle|^2.$$

This is the smooth, nonfluctuating part of the scattering, sometimes called the shape elastic. Since $\bar{\sigma}_{ce} = \sigma_{fl}$ and $\sigma_{el}^{op} = \bar{\sigma}_{el} - \bar{\sigma}_{ce}$, we again find that this shape-elastic scattering corresponds to the scattering of the model (i.e., $\sigma_{el}^{op} \doteq \frac{\pi}{k^2} |1 - \langle \eta \rangle|^2$). The optical model is therefore connected to the averaged problem in which the fluctuations are removed entirely by transferring them from scattering to absorption:

$$\sigma_{el}^{op} = \bar{\sigma}_{el} - \bar{\sigma}_{ce} \text{ and } \sigma_{a}^{op} = \bar{\sigma}_{r} + \bar{\sigma}_{ce}.$$

The validity of this correspondence depends on the equality of fluctuation with compound elastic scattering.

The crucial point in our argument is the recognition that $\bar{\sigma}_{fl} = \bar{\sigma}_{ce}$. More insight into this equation is provided by examining the time behavior of a neutron wave packet. The fluctuation scattering of the wave packet comes out late compared to the rest, thus identifying itself with the part of the wave function which was delayed by forming the compound states. Basically the difference in time behavior of the scattering associated with σ_{se} and with σ_{fl} can be seen immediately from the definitions. σ_{se} was designed to vary slowly with the slow-varying $\langle\eta\rangle$; σ_{fl} had to take up the rapid η variations omitted from σ_{se}. The shape-elastic scattering must have a time behavior more or less like a wave packet in the old single-particle model. The fluctuation elastic scattering on the other hand involves $\Delta\eta = \eta - \langle\eta\rangle$, and the behavior of $\Delta\eta$ near a compound resonance brings the width $\Gamma^{(s)}$ into the time behavior of the reflected packet.

Explicitly, we write (6) in the form

$$\psi = \frac{A}{r}(e^{-ikr} - \langle\eta\rangle e^{ikr}) - \frac{A}{r}(\eta - \langle\eta\rangle)e^{ikr} \qquad \text{for } r > R. \qquad (15)$$

In this way we put the interaction described by the average phase into the first term, and the fluctuations of η show up in the second. Dropping the last term would give a model in which the scattering is σ_{se} and the absorption σ_c: It would leave only the part we have identified with the optical model.

In conformity with our average over an energy interval I great compared with the level spacing D, we now build a wave packet out of phase waves (15), the incoming part of which passes a given point in a time $T \sim \hbar/I$. We can then examine the time behavior of the outgoing parts of the wave packet as they pass a given radius, looking separately at the shape elastic term

$\frac{A}{r}\langle\eta\rangle e^{ikr}$ and at $\frac{A}{r}(\eta-\langle\eta\rangle)e^{ikr}$. Since the average value $\langle\eta\rangle$ is constant over the energy interval of the pulse, the scattered pulse corresponding to the first of these two terms has the same shape as the incident pulse and will emerge immediately as the incident pulse sweeps over the nucleus.

The second term, corresponding to the fluctuation, is the more interesting one. Its time dependence is given essentially[4] by

$$f(t) = \int_I (\eta-\langle\eta\rangle)e^{-iEt/\hbar}dE.$$

On putting in for η the approximate Breit-Wigner η_{BW} of Eq. (13), we can evaluate this time behavior. For $t > T$, we obtain

$$|f(t)|^2 = \sum_{s,s'} \Gamma_n^{(s)}\,\Gamma_n^{(s')}\,e^{i(E_s-E_{s'})t/\hbar}\,e^{-\frac{\Gamma^{(s)}+\Gamma^{(s')}}{2\hbar}t}. \tag{16}$$

In (16) all the periods $\frac{2\hbar}{\Gamma^{(s)}+\Gamma^{(s')}}$ are of the order $\hbar/\Gamma^s = \tau_c^s$. Hence $|f(t)|^2$ decays in a time of the order τ_c^s.

We may recall the inequality $\hbar/\Gamma^{(s)} \gg \hbar/D \gg \hbar/I$. (That is, the lifetimes τ_c^s of the compound system are long compared to the internal repetition time τ, which, in turn, is much greater than the length T of the wave packet.) Consequently, $\tau_c^s \gg T$. We see therefore that the emission of the fluctuation wave packet takes place almost entirely after the time T, that is, almost without any interference with the shape elastic scattering. As expected, the fluctuation gives elastic scattering which is delayed and

[4]The wave packet is observed going in and out reasonably close to the nucleus so that it has no time to spread during its motion.

which arrives (according to (16)) with the decay periods of the compound states. Our interpretation of σ_{fl} as σ_{ce} is reinforced by the time behavior.[5]

We can now summarize the progress of a nuclear reaction as follows: When the incident wave packet reaches the nucleus, a large part of the pulse is scattered by the nuclear potential well, and a scattered pulse of roughly the same form as the initial one leaves the nucleus at once. At the same time—that is, within the coalescence time $\hbar/V_1$ on the optical model—part of the incident pulse forms a compound nucleus. Because this compound forming part returns to the channel entrance only after the time τ, it has no influence upon the original scattered pulse.

[5] In (16) we may split the sum into $\sum_s (\Gamma_n^{(s)})^2 e^{-\Gamma^{(s)}t/\hbar} + \sum'_{s'<s}$ where $\sum'$ contains all the cross terms between the various resonances. In addition to the exponential decay, each cross term contains the oscillatory time factor $\cos (E_s - E_{s'})t/\hbar$. In special circumstances, these terms may interfere at certain times to make a large contribution to $|f(t)|^2$. On the time average, however, they are small compared to the diagonal ($s' = s$) sum as long as $\Gamma \ll D$. Crudely speaking, therefore,

$$|f(t)|^2 \sim \sum_s (\Gamma_n^{(s)})^2 e^{-\Gamma^{(s)}t/\hbar}$$

i.e., $|f(t)|^2$ shows the decays of the separate compound states.

It is interesting to note that as a function of energy the compound elastic wave packet stretches over the energy interval I, but it is concentrated, as one would expect, in narrow energy bands at the energies of the resonances of the compound system. This concentration is related to the long times τ_c^s during which the compound elastic scattering comes out.

We also remark (still for $\Gamma/D \ll 1$) that in

$$\frac{1}{I}\int |f(t)|^2\, dt = C\left(\sum \frac{(\Gamma_n^{(s)})^2}{\Gamma^{(s)}} + \int \Sigma'\, dt\right)$$

the second term is negligible, so that

$$\frac{1}{I}\int |f(t)|^2\, dt \sim \frac{1}{I}\sum \frac{(\Gamma_n^{(s)})^2}{\Gamma^{(s)}}$$

in agreement with (14) for $\bar{\sigma}_{ce}$.

Here we are using the fact, discussed on page 140, that $\tau = 2\pi\hbar/D$ represents a "time of revolution" in the compound state; it is of the order of the time it takes for the pulse to "reappear" at the entrance channel after having produced a compound state. As $\tau \gg T$, the reappearance at the entrance channel is long after the original scattering has taken place. Consequently the original scattering of the pulse cannot be altered by the establishment of any special boundary conditions, or, in other words, the internal pulse which forms the compound system cannot interfere with the original pulse to modify the immediate reflection. Also the re-emission of the particle by the compound nucleus into the entrance channel will take place much later, after the time $\hbar/\Gamma$, and can therefore be distinguished from the scattering that takes place immediately. Hence, for a pulsed initial neutron beam, the compound formation can be clearly separated and appears as an absorption of part of the incident pulse in spite of the fact that re-emission may occur later on. Finally, turning the argument around, since the pulse time T must be short compared to the repetition time $\tau = 2\pi\hbar/D$, the energy spread of the incident beam necessarily must be large compared to D. The cross sections which are defined by this consideration therefore must be averages over the resonances of the compound nucleus.

7. The predictions of the complex potential model compare fairly well with the experimental results. We shall classify the experiments into the following three groups: total cross sections, elastic scattering data, and cross sections for the formation of the compound nucleus.

Total cross sections. As indicated before, the prediction of the total neutron cross sections shows a surprising agreement with the measurements, even when a simple square well is used for the potential. This agreement is especially significant in the low energy region, between 0 and 2 Mev, in which the total cross section, when averaged over individual resonances, exhibits pronounced maxima and minima. These characteristic features are very well represented by the calculations when the constants (5) listed above are used. The calculation is, moreover, reasonably sensitive to changes in V_o, V_1, and R, so that the fit determines their values rather precisely [25].

At somewhat higher energies the total cross sections become less sensitive to the constants. They are roughly approximated [8] by the formula $\sigma_{tot} = 2\pi(R+\lambda\!\!\!^{-})^2$ and have very little dependence on the imaginary part of the potential. Nevertheless, the observed deviations from this form might be used to get some information regarding the constants of the potential well. At present the main obstacle to a determination of V_1 above 4 Mev is of a mathematical nature. It is difficult to calculate the scattering by a potential well whose dimensions are large compared to the wave length, and whose depth is too big for the use of the Born approximation.

For scattering at much higher energies the first application of the optical model was done by Fernbach, Serber, and Taylor [24]. This important work was actually the first systematic attempt to represent the effect of the nucleus by a complex potential. Recently T. B. Taylor [27] has extended and refined the method and has applied it to the scattering from 40 Mev up. He finds that the experimental results near 40 Mev can be reproduced by a potential well which is slightly less deep than (5); also the depth of the well must decrease further with increasing energy. The imaginary part of the potential is considerably larger than the one found at low energy, but it also decreases towards still higher energies. (See the Additional Remarks at the end of this paper.)

Elastic scattering. The scattering cross section σ_{el}^{op} and the absorption cross section σ_a^{op} calculated on the basis of the complex potential model are somewhat more difficult to compare with the experimental data. The scattering on the optical model does not include the compound elastic scattering which plays an important role at energies below 1 or 2 Mev. Similarly, the calculated absorption represents the formation of the compound nucleus, and it too has not been measured directly, since it contains all possible reactions including the compound elastic scattering.

By estimating the compound elastic scattering, however, some conclusions can be drawn from presently available experiments. With a plausible estimate of Γ_n/Γ (i.e., of σ_{ce}/σ_c), the same simple complex potential which represents the total cross section between 0 and 3 Mev also reproduces the observed angular dependence of the scattering at 1 Mev [28], [25].

At higher energies, where the compound elastic scattering is suppressed by competing processes, the observed scattering can be compared directly with the model. Recent measurements of the angular dependence of the elastic scattering of neutrons at 4.1 Mev by Walt and Beyster [29] can be represented fairly well by a complex potential well with the same values of V_o and R used before. The imaginary part, V_1, must be increased by at least a factor 3, which indicates an increased probability of compound nucleus formation at this increased energy.

If the neutron energy is raised still higher, the general pattern of the angular distribution no longer depends sensitively on the details of the potential well other than the radius. The dependence on radius is represented by the well-known diffraction pattern of a circular disc. Only a very refined study would provide any further information about the potential well.

So far such studies have only been made with protons. For several elements scattering of 10 Mev protons [30] has been measured, and the results for oxygen have been analyzed on the optical model. They are well represented by $V_o = 30$ Mev and $V_1 = 5$ Mev [31]. Also the scattering of protons of approximately

20 Mev has been measured [32]; and Saxon and Wood [33] have attempted to explain the measurements in terms of the optical model. They are successful in reproducing the main features of the experiment by increasing the imaginary part to 10 Mev and by rounding off the edges of the square well. The region of rounding in which the potential rises from the value $-V_0$ to zero has to be made about 1×10^{-13} cm wide. This rounding off is a very natural change, and it should be expected to be of importance at higher energy and for scattering at large angles.

Compound nucleus formation. The difficulties of comparing the theoretical predictions of the cross section σ_c for the formation of the compound nucleus with experimental results can be overcome in the following way. At low energies the predicted values of σ_c are supposed to be average values over resonances; consequently, we can make use of the fact that the resonances are well represented by the Breit-Wigner formula. We then obtain the following relation for σ_c in the case of very low energies (neutrons of $l = 0$ only):

$$\sigma_c = 2\pi^2 \lambda\!\!\!^{-2} \left(\frac{\overline{\Gamma}_n}{D}\right) = \frac{C(A)}{v}$$

where $\left(\frac{\overline{\Gamma}_n}{D}\right)$ is the average of the neutron width over the level distance taken for neighboring levels, and $C(A)$ is a function of the atomic number. This expression allows a direct check of the predictions for σ_c at low energies. (See the Additional Remarks at the end of this paper.)

It is easy to see that our model predicts maxima for $v\sigma_c = C(A)$ at values for which a standing wave can develop within the nucleus. This is the case when $\sqrt{2mV_0}\,R = \pi\hbar(n + \frac{1}{2})$, where n is an integer. Hence, within the range of nuclear radii available, we should expect a maximum for $v\sigma_c$ and also for $\left(\frac{\Gamma_n}{D}\right)$ near the mass numbers $A \sim 11$, 55, and 155. The maximum at $A \sim 155$ has been established by Carter, Harvey, Hughes, and Pilcher [34]. It is not quite as high as the optical model would predict with the constants established by the scattering experiments. However, fluctuations in the A-dependence of the radii and strong deviations from sphericity might cause a flattening of the expected maximum. At $A \sim 55$ and 11 the level spacing is large, and it is more difficult to measure the constants of several levels in order to obtain a valid average. Quite recently, however, a maximum at $A \sim 55$ was found by Cote and Bollinger [35].

The predictions of the optical model in respect to σ_c can also be checked by comparing them with the observed reaction neutron cross sections, σ_r. The cross section σ_c must be larger than σ_r, and the difference must be accounted for by compound elastic scattering. The quantitative agreement with the results derived from the square well potential is somewhat less satisfactory than the agreement of the total neutron cross sections.

At 1 Mev the experimental values exhibit pronounced maxima and minima as a function of A [28]. The theory also shows maxima and minima, but with the values of R used above they occur at wrong values of A for the heavy elements [25]. By using smaller radii for these elements [36], this discrepancy can be removed at least to within present experimental accuracy.[6] (The necessary modification of R does not shift the peak in $v\sigma_c$ too far and even improves the fit to σ_{tot}.) At higher energy the maxima and minima are less pronounced both experimentally and

[6] It may be noted that the comparison of the theory and the experiments concerning the reaction cross section in this energy range severely tests them both. The reaction cross section is measured as a difference between the total cross section and the integrated differential elastic cross section. This difference is only $\sim \frac{1}{6}$ of the total.

theoretically; however, the experimental cross sections are somewhat smaller in general than the theoretical prediction given by the square well potential [29]. Rounding the edge of the potential well will increase the theoretical values.

This application of the complex potential model to the elastic scattering and to the compound nucleus formation is in some respects a more incisive test than the agreement of the calculations with total cross sections. Here the agreement depends on the details of transferring the fluctuation from one cross section to another, whereas the total cross section, depending directly on $\langle \eta \rangle$, is completely independent of this split of absorption and scattering. As the test here is somewhat more sensitive than that provided by total cross sections, it is not too surprising that the agreement is somewhat less good.

8. In this essay we have tried to show that a more careful application of Bohr's original ideas and an analysis of the vastly increased experimental material leads to a modification of the primitive picture previously used to describe nuclear reactions: According to the old picture, the incident particle hits the target nucleus and forms a compound system in which its energy is shared among all constituents. The compound system decays into the reaction products in a way which is independent of the process of formation.

Now this view must be modified in several ways. The formation of a compound system does not necessarily occur whenever the incident particle penetrates the nucleus. In fact, the effect of the target nucleus upon the particle can be described reasonably well (when averaged over resonances) by a complex potential. Part of the effect is simply a scattering in which the target nucleus acts as a potential well only. The other part is the compound nucleus formation, which occurs with a much smaller probability than previously anticipated. Here we understand by compound formation any process in which the incident particle is removed from the entrance channel. It includes not only the processes in which the incident nucleon shares its energy with the whole nucleus, but also energy transfers to one or to a few constituents of the target nucleus.

The decay of the compound system into the reaction products depends upon the detailed mechanism of the energy transfer. The decay is independent of the mode of formation only in certain limiting cases. In general, the reaction products and their energetic and angular distribution will depend on the special conditions that prevail when the compound system is formed. In order to understand, classify, and calculate the different mechanisms that come into play in nuclear reactions, a detailed study will be necessary of the interactions between the incident particle and the individual and collective motions of the nuclear constituents. (See the Additional Remarks at the end of this paper.)

Bohr's compound nucleus picture provided insight into many phenomena; it demonstrated its "peculiar facilities for a comprehensive interpretation of the characteristic properties of nuclei in allowing a division of nuclear reactions into well-separated stages to an extent which has no simple parallel in the mechanical behavior of atoms." [1]. Now after two decades our knowledge extends beyond the limits of this description.

Remarks Added by the Author, 1972.

Remark to p. 147. Today there are many more experimental results available from which one can determine the average potential in the optical model. The assumption (5) of a square well potential, identical for protons and neutrons, turned out to be approximately correct, but too simple. Today one uses potentials with rounded edges, slightly different for neutrons and protons, and one adds a spin-orbit coupling. Most importantly, the experimental results are much better reproduced when the imaginary part of the potential is assumed to be proportional not to the real part but to the derivative of the real part. It localizes most of the absorption at the surface. This is plausible from the point of view expressed in the article "Nuclear Models" of this collection. At the surface the density is smaller, not all states below the Fermi energy are filled, and hence collisions can take place with higher probability. In contrast to the statement on p. 160, the energy dependence of the imaginary part of the potential is not very strong. For a more quantitative analysis of the potential, see *Nuclear Structure Studies with Neutrons*, edited by Neve de Mevergnies, Van Assche, and Vervier, North Holland Publishing Company, 1965.

Remark to p. 163. The dependence on A of $v\sigma_c = C(A)$ is now well established and explained by the optical model. The deviations from sphericity are known today, and they explain not only the flattening of the maximum at $A \sim 155$, but also its split into two maxima. The limitations of the optical model account for the fact that within the minima between the maxima some structure is found. One can explain this structure by the fact that after the first collision of the entering nucleon, there will be two nucleons moving, each of them preferring to be in a resonance. (see "Problems of Nuclear Structure".) Those secondary two-particle states are called "doorway states."

Remark to p. 165. The experience after 1954 has shown that Bohr's compound-nucleus picture is more useful and applicable than the remarks at the end of this article imply. The arguments of this article have shown that the establishment of the compound nucleus is more subtle than in the original Bohr assumption: The incoming particle moves first almost like a free particle in the optical potential, but after it collides with other nucleons and those collide again a few times, the original energy is soon dissipated among many constituents of the nucleus and a true compound state is established whose properties are not much dependent on the special conditions when it was formed. Therefore, apart from the formation probability, most of Bohr's conclusions about the compound nucleus, in particular its decay, remain valid today. For example, we know today many instances where the energy distribution of the secondary neutrons emerging from a nuclear reaction is truly Maxwellian.

References

[1] N. Bohr, *Nature* **137,** 344, 1936.
[2] F. Fermi *et al., Ric. Sci.* **5,** 282, 1934.
[3] H. A. Bethe, *Phys. Rev.* **47,** 747, 1935; F. Perrin and W. Elsasser, *J. de Phys.* **6,** 195, 1935; F. Fermi *et al., Proc. Roy. Soc.* A **149,** 522, 1935; G. Beck and L. H. Horsley, *Phys. Rev.* **47,** 510, 1935.
[4] T. Bjerge and C. H. Westcott, *Proc. Roy. Soc.* A **150,** 709, 1935; P. B. Moon and Tilman, *Nature* **135,** 904, 1935; L. Szilard, *Nature* **136,** 849 and 950, 1935; F. Fermi and E. Amaldi, *Ric. Sci.* A **6,** 544, 1935; R. Frisch *et al., Nature* **137,** 149, 1936; R. Frisch and G. Placzek, *Nature* **137,** 357, 1936; P. Preiswerk and H. Halban, *Nature* **138,** 163, 1936.
[5] G. Breit and E. P. Wigner, *Phys. Rev.* **49,** 519, 1936.
[6] H. A. Bethe and G. Placzek, *Phys. Rev.* **51,** 450, 1937.
[7] L. Eisenbud and E. P. Wigner, *Proc. Nat. Acad.* (*U.S.A.*) **27,** 281, 1941; E. P. Wigner and L. Eisenbud, *Phys. Rev.* **72,** 29, 1947; T. Teichman, *Phys. Rev.* **77,** 506, 1951.
[8] H. Feshbach and V. Weisskopf, *Phys. Rev.* **76,** 1550, 1949.
[9] I. Frenkel, *Sov. Phys.* **9,** 533, 1936; V. Weisskopf, *Phys. Rev.* **52,** 295, 1937; L. Landau, *Phys. Zeits. Sow.* **11,** 556, 1937.
[10] J. P. Blaser *et al., Helv. Phys. Acta.* **24,** 3, 1951.
[11] P. C. Gugelot, *Phys. Rev.* **81,** 51, 1951; E. R. Graves and L. Rosen, *Phys. Rev.* **89,** 343, 1953.
[12] H. Barschall, *Phys. Rev.* **86,** 431, 1952; D. W. Miller *et al., Phys. Rev.* **88,** 83, 1952; M. Walt *et al., Phys. Rev.* **89,** 1271, 1953; A. Okazaki, S. F. Darden, and R. B. Walton, *Phys. Rev.* **93,** 461, 1954; N. Nereson and S. F. Darden, *Phys. Rev.* **89,** 775, 1953 and *Phys. Rev.* **94,** 1678, 1954; C. F. Cook and T. W. Bonner, *Phys. Rev.* **94,** 651, 1954.
[13] E. B. Gugelot, *Phys. Rev.* **93,** 425, 1954; E. B. Paul and R. L. Clark, *Can. J. Phys.* **31,** 267, 1953; R. Eisberg and G. W. Igo, *Phys. Rev.* **93,** 1039, 1954; H. McManus and W. T. Sharp, *Phys. Rev.* **87,** 188, 1952.
[14] J. M. Blatt and V. Weisskopf, *Theoretical Nuclear Physics,* Ch. IX, § 5.
[15] S. N. Ghoshal, *Phys. Rev.* **80,** 939, 1950.

[16] B. Cohen, *Phys. Rev.* **92,** 1245, 1953.
[17] N. Bohr, Peierls, and G. Placzek, *Nature* **144,** 200, 1939.
[18] G. Bernardini, S. J. Booth, and E. T. Lindenbaum, *Phys. Rev.* **88,** 1017, 1952; M. Goldberger, *Phys. Rev.* **74,** 1269, 1948.
[19] H. A. Bethe, *Phys. Rev.* **53,** 675, 1938.
[20] V. Weisskopf, *Helv. Phys. Acta* **23,** 187, 1950.
[21] S. Drell and K. Huang, *Phys. Rev.* **91,** 1527, 1953; K. A. Brueckner, C. A. Levinson, and H. M. Mamhoud, *Phys. Rev.* **95,** 217, 1954.
[22] M. H. Johnson and E. Teller, *Phys. Rev.* **93,** 357, 1954.
[23] V. Weisskopf, *Science* **113,** 101, 1951.
[24] H. A. Bethe, *Phys. Rev.* **57,** 1125, 1940; S. Fernbach, R. Serber, and T. B. Taylor, *Phys. Rev.* **75,** 1352, 1949.
[25] H. Feshbach, C. Porter, and V. Weisskopf, *Phys. Rev.* **96,** 448, 1954; R. K. Adair, *Phys. Rev.* **94,** 737, 1953; H. Feshbach, C. Porter, and V. Weisskopf, *Phys. Rev.* **90,** 166, 1953.
[26] J. M. Blatt and V. Weisskopf, *Theoretical Nuclear Physics.*
[27] T. B. Taylor, *Phys. Rev.* **92,** 831, 1953.
[28] M. Walt and H. Barschall, *Phys. Rev.* **93,** 1062, 1954.
[29] M. Walt and J. R. Beyster, *Phys. Rev.* **48,** 677, 1955.
[30] W. E. Burcham, W. M. Gibson, A. Hossain, and J. Rotblat, *Phys. Rev.* **92,** 1266, 1953.
[31] Prowse and A. Hossain (Bristol), private communication.
[32] J. W. Burkig and B. T. Wright, *Phys. Rev.* **82,** 451, 1951; B. L. Cohen and R. V. Neidigh, *Phys. Rev.* **93,** 282, 1954; R. C. Gugelot, *Phys. Rev.* **93,** 425, 1954.
[33] D. S. Saxon and R. D. Wood, *Phys. Rev.* **95,** 577, 1954.
[34] R. S. Carter *et al.*, *Phys. Rev.* **96,** 113, 1954.
[35] R. Cote and Bollinger, *Bull. Amer. Phys. Soc.* **30,** (No. 1), 23, 1955.
[36] W. S. Emmerich, personal communication.
[37] H. Gittings, H. H. Barschall, and G. G. Everhart, *Phys. Rev.* **75,** 610, 1949; D. Phillips, R. W. Davis, and E. R. Graves, *Phys. Rev.* **88,** 600, 1952.

Problems of Nuclear Structure

Only six and a half years ago an international conference on nuclear physics was held in Glasgow in which both the physics of elementary particles and the physics of nuclear structure were discussed. It was the last conference in which these two fields of physics were considered as one and the same. Since then, nuclear physics has split into two definite branches, one dealing with the nature and the properties of elementary particles and the other dealing with the structure and the dynamics of atomic nuclei. Today these two fields are as far apart as solid-state physics and nuclear physics. The specialists in one field know very little about what is going on in the other field. This is a very deplorable state of affairs since both fields are full of exciting developments.

This survey is devoted to the second field, the study of nuclear structure. Its object is the investigation of the properties of atomic nuclei as systems of neutrons and protons in close contact. It deals with the behavior of these particles under the influence of the nuclear forces which act between them. Since the great discovery of Yukawa we know that the nuclear forces are carried by mesons. Hence the complete understanding of nuclear phenomena would imply a description of the nucleus as a system of nucleons in interaction with a common meson field—nucleons that are not identified as protons or neutrons, swimming in a sea of virtual mesons. Fortunately, and somewhat surprisingly, the nucleus can be rather accurately described as a system of well-defined neutrons and protons with certain forces between them. The meson origin of these forces does not seem to play an essential role in the nuclear behavior at lower energy. Hence the theory of nuclear structure is not interested in the theory of the nuclear force itself; it is taken for granted

Revision by the author of an article originally in *Phys. Today* **14,** 7, 18, July 1961.

that such a force exists and its properties are accepted as an experimental fact.

Let us begin with a description of nuclear forces as far as we know them today. One assumes that the nuclear force can be represented by a potential energy which depends upon the distance between the nucleons, their relative spins, and symmetry properties. We assume today that the nuclear force acts only between pairs of particles; that means that the nuclear potential is a sum of terms, each coming from the interaction of a pair of nucleons. This is an assumption, which is made mainly because of its simplicity. So far there is no experimental evidence which forces us to abandon it.

The nuclear potential can be divided into three parts: The first one is a central force, the second is the well-known tensor force, and the third is a spin-orbit force:

$$V = V^C(r)+S_{12}V^T(r)+(l\cdot s)V^{LS}(r). \tag{1}$$

Here S_{12} is the characteristic tensor force term, $S_{12} = 3(r\cdot s_1)(r\cdot s_2)/r^2-(s_1\cdot s_2)$; s_1 and s_2 are the spins of the two nucleons, $s = s_1+s_2$; and l is the relative angular momentum. The three functions $V(r)$ depend on the distance, the relative spin, and the relative symmetry of the two particles. The dependence on the distance of the first two forces is shown in Fig. 1. There is a repulsive core with a radius c of about 0.4 f, outside of which there is an attractive potential with a range of 1 to 2 f. The spin-orbit potential has a shorter range and plays a minor role in most problems of nuclear physics.

The main forces, the central and tensor forces, are strongly symmetry dependent. They are active almost exclusively in even states. The odd states show very little nuclear force. This property is usually referred to as the characteristic of a Serber force. The spin dependence of the central force is relatively weak. It

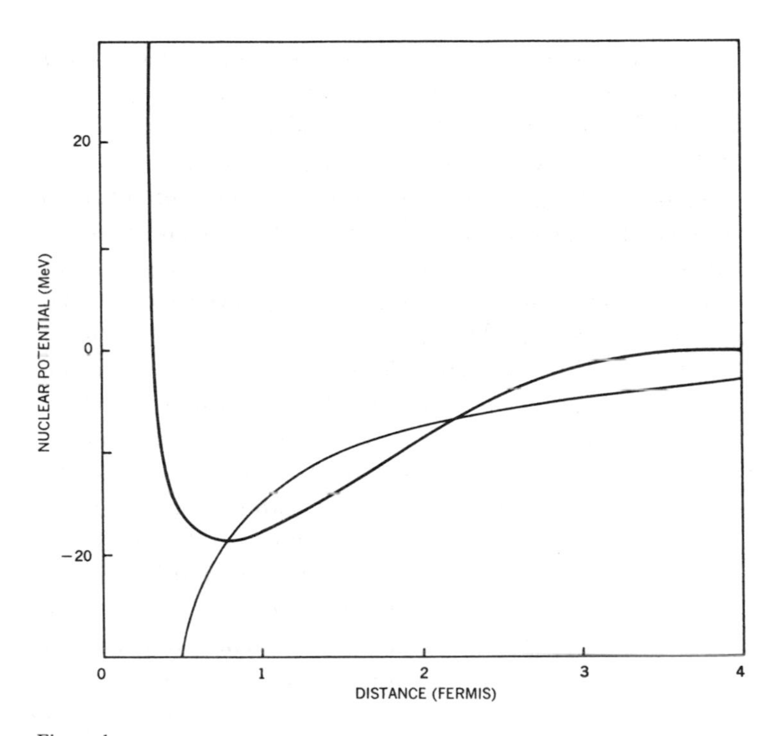

Figure 1
Dependence on distance of central and tensor forces. The potential function $V(r)$ is plotted against the distance in fermis. Ordinate is in arbitrary units; core radius is denoted by c.

can be disregarded in qualitative considerations. The tensor and spin-orbit forces act only in triplet states, of course.

Once the nuclear forces are given in this form, all problems of nuclear structure are in principle problems of solving a Schrödinger equation for A particles. Evidently such an equation cannot be solved in general. Hence one has to resort to approximative methods, to nuclear models, and to other devices for drawing conclusions from the consequences of nuclear interaction. There are, however, a number of fundamental problems which can be solved to some extent by solving the Schrödinger equation itself. These problems are a direct test for our assumptions concerning the nuclear force. The most important ones are: (1) the deuteron; (2) the scattering of nucleons by nucleons as a function of energy and polarization; (3) the treatment of the three-body systems, He^3, H^3; and (4) the problem of nuclear matter.

Obviously only problems 1 and 2 can be solved exactly since they are two-body problems. In fact, they have served so far as the only source of information for the determination of the nuclear forces as given above. From the study of these two problems one has concluded that it is possible to describe the nuclear force by a velocity-independent potential of the type (1). It should be emphasized, however, that this determination is not unique; it certainly would be possible to represent the facts of the problems (1) and (2) by other potentials which depend on the velocity.

Problems (3) and (4) can only be solved by approximation, but a high degree of approximation can be reached, in particular for problem (3). According to the latest investigations by Derrick, Mustard and Blatt [1] one still has not found a potential of form (1) which is good for problems (1) and (2), and also gives the right energy and shape for the three-body systems. It is not proven yet that this task is impossible; in fact it is highly probable that such a two-body potential exists and that no three-body force is necessary for the description of the three-body systems.

It is perhaps worthwhile to treat the problem of nuclear matter in some more detail. Nuclear matter is the hypothetical substance which one would get if the electric repulsion between protons were eliminated. The analysis of the properties of nuclei makes it extremely plausible that, under these conditions, an infinite number of neutrons and protons would form, as a state of lowest energy, a substance with the following properties: The binding energy per particle would be 15 Mev and its density ρ, as given by the average distance between neighbors, would be $d = 1.8$ f ($d = \rho^{-\frac{1}{3}}$). The dynamics of nuclear matter would be such that the nucleons would move within it almost as free particles.

Brueckner and collaborators were the first who were able to derive these properties of nuclear matter directly from the nuclear forces. Many other authors have since reformulated and sometimes improved this calculation. The main idea is based on the fact that the nuclear forces are in reality quite weak, as indicated by the small binding energy of the deuteron. Therefore, in an assembly of protons and neutrons, the Pauli principle is sufficient to prevent the nuclear forces from having a strong distorting effect upon the wave functions of the particles. Hence one is justified to start out to describe nuclear matter in terms of a Fermi gas of free particles and treat the effect of the nuclear attraction as a perturbation. The only restriction in this approach comes from the repulsive core. Although the core radius is rather small compared to the average distance of nucleons (it is as small as the molecular radius would be to the average distance of molecules in a gas of 12 atmospheres), it still has a characteristic influence upon the energy of a Fermi gas. This influence can be expressed in a qualitative way by the following change in the expression of the kinetic energy per particle $E_K = B/d^2$ of a free Fermi gas:

$$E_K = \frac{B}{(d-c)^2}, \qquad B = \tfrac{3}{10}\left(\tfrac{3}{2}\pi^2\right)^{\frac{2}{3}} h^2/m = 27 d_0{}^2 \text{ Mev.}$$

The only change is the replacement of d by $d-c$. (d_0 is the value of d in actual nuclear matter: $d_0 = 1.8$ f.)

The potential energy E_P of the remaining attractive forces can be expressed by the following simple expression, keeping in mind that nuclear matter is almost a gas of free particles:

$$E_P = \frac{C}{d^3} f(d).$$

If there were no correlations whatsoever between the particles, and if the central forces were not symmetry dependent (the tensor and spin-orbit forces do not contribute in first order in a free gas), $f(d)$ would be unity and $C \sim (4\pi/3)\ V_0 b^3$, where V_0 is the depth and b the range of the central forces. Actually, however, $f(d)$ is not unity for two reasons:

a. The central forces act only in even states and the Fermi statistics forbid equal particles of equal spin to be in even states.

b. The attractive part of the nuclear forces will tend to keep pairs together and establish correlations, thus increasing the effect of the central force and allowing the tensor force to operate.

The effect of these two points is shown qualitatively in Fig. 2. The dotted curve is the effect of point (a); the solid curve is the combined effect of point (a) and point (b). The influence of

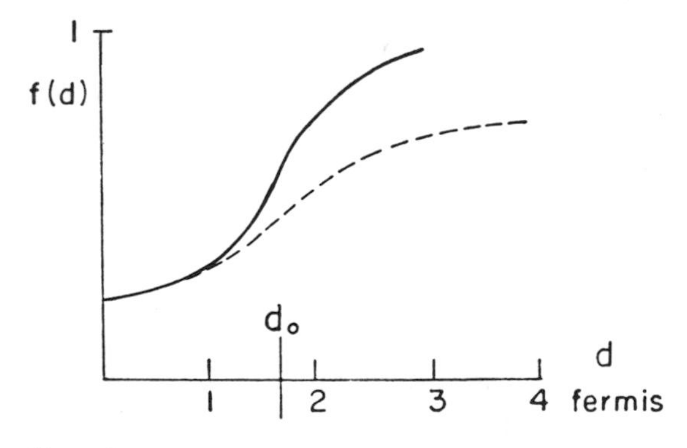

Figure 2.
Correction function for particle correlations in nuclear matter; $f(d)$ is plotted against average distance between particles in fermis, while d_0 is equilibrium distance. Dotted curve corresponds to the Born approximation.

point (b) is kept small by the rigidity of the Fermi distribution against the effect of forces, in particular when the Fermi energy is high, as it is for small values of d. Near $d = d_0$ one can approximate $f(d)$ by $f(d) \sim \frac{1}{2}(d/d_0)$.

We can now write down an "equation of state" for nuclear matter. The energy ϵ per particle is very approximately

$$\epsilon \cong \frac{B}{(d-c)^2} - \frac{C}{d^3} f(d).$$

One can find the equilibrium density and energy by setting zero the derivative of ϵ with respect to d. The stability of nuclear matter at low densities ($d_0 \sim 4.5\ c$) is seen to result from the increase of the kinetic energy because of the repulsive core, and because of the weakness of the nuclear forces [$f(d_0) \sim \frac{1}{2}$] coming from the absence of forces in the odd states—a weakness which is not overcome by correlations favoring the attractive forces and the tensor forces, since the Fermi distribution resists any appreciable change of wave functions away from those of the free Fermi gas.

The equation of state also permits calculating the compressibility $\frac{1}{2}d^2(\partial^2\epsilon/\partial d^2)$ of nuclear matter which gives a result of 100 Mev, and the symmetry energy E_S: $a_S = \frac{1}{2}(\partial^2\epsilon/\partial y^2)$ with $y = (N-Z)/(N+Z)$, where N and Z are the numbers of neutrons and protons. a_S can be estimated by these methods to be between 25 and 30 Mev.

An interesting question has recently been posed in this connection: Does a large assembly of neutrons form stable nuclear matter? This problem is important for astrophysicists, who believe that such a form, if stable, might occur in the center of stars at the end of their development. Recent investigations by Salpeter [2] leave the question still open. The nuclear forces are not known with sufficient accuracy to answer this question theoretically. This question received a positive answer with the recent discovery of pulsars. Today the evidence seems convincing that pulsars are in fact "neutron stars." Their interiors consist of neutron matter. A large assembly of neutrons does indeed form stable nuclear matter, compressed by gravity to the immense density of 10^{15} g/cm^3 (one cubic centimeter weighs a billion tons), a density at which the neutrons almost "touch" each other, if we endow them with a radius of about one fermi.

Most problems of nuclear structure cannot be dealt with by direct approach in solving the Schrödinger equation for all nuclear particles involved. In all but very few problems a completely different policy must be employed. One uses as little as possible of our detailed information of nuclear forces and tries to concentrate on qualitative features. New concepts and pictures are introduced, such as radius, potential well, surface, shape, which can be defined and measured empirically. They play the same role as the material constants in solid-state physics, such as density, elasticity, viscosity, shearing coefficient, etc. They are useful for the description of the observed facts, but their connection with the fundamental forces is vague and only qualitatively understood.

This policy of dealing with nuclear facts brought about the introduction of nuclear models of many kinds: the independent-particle model, the optical model, the collective model, the statistical model, and many more. One sometimes had the impression of using contradictory hypotheses for the explanation of different data. This is not so. A model is nothing else but the overstressing of certain features which are responsible for the phenomenon under consideration. We propose to show that all the models in use today stem from the same ideas and fit together logically. Far from contradicting one another, they are different aspects of the same principle, which follows from our study of nuclear matter, the principle that the motions of particles in the nucleus in first approximation are independent ones, and that the effect of the interaction can be considered as a perturbation in the next approximation. Thus modified and expanded, the independent-particle picture can be used as a basis for the understanding of many nuclear phenomena. This is not to say that everything is understood and explained. On the contrary, the concepts and methods used are very qualitative and only those phenomena can be explained which are connected with very general properties of the forces. The nuclear surface, for example, its structure and effects, is known only vaguely and empirically without much basic explanation; so are many other features of the nucleus, which depend on the more detailed nature of the nuclear forces.

We start our description of nuclear behavior with the Independent Particle Model (IPM). It describes the nucleus as a spherical potential well, in which the nucleons move as non-interacting free particles. The well has the size of the nucleus; its depth is somewhat momentum dependent, and contains a spin-orbit term. This common potential is the average effect of all nucleons on a single one, the spin-orbit term coming from the original spin-orbit force between nucleons and also from the overall effect of the tensor force. Such a potential well gives rise to the well-known shell structure. Whenever the number of protons or neutrons reaches a "magic" value, we obtain a closed and stable ensemble of particles.

We therefore describe an average nucleus as consisting of a "core" and a "cloud." The core contains the particles in closed shells and the cloud consists of the n nucleons in open shells; n can also be negative, in which case we are dealing with a certain number of holes in a closed shell. This division will help us to describe the properties of nuclei, in particular their spectra.

The simplest kind of spectrum is that of nuclei whose clouds consist of one particle or hole, $n = \pm 1$. We then expect a single-particle spectrum, exhibiting the levels in the next open shell or the spectrum of a hole in the last closed one. Figure 3 shows Pb^{207} as an example.

As the next step, we consider a nucleus with a larger cloud. The interaction between nucleons in open shells has characteristic effects upon the dynamics. According to the Copenhagen school, one divides the nuclear forces into two parts, and distinguishes short-range "pairing" forces and long-range "quadrupole" forces. The former bind the nucleons into pairs of zero angular momentum. If they were to predominate in the cloud, it would have a spherical shape. The latter are forces which favor a good overlap of wave functions among all *n* particles. If they predominate, they favor an arrangement in which all particles concerned are concentrated into certain directions and not in other directions. Hence they produce a deformed cloud.

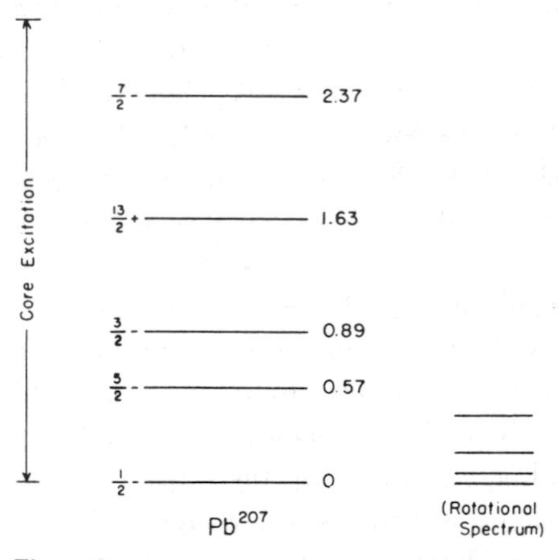

Figure 3.
Single-particle spectrum in the field of a core. (Core ± 1.) The present example is the spectrum of a hole. For purposes of comparison, the first core excitation is indicated at left and a rotational spectrum at right.

The former forces produce saturated pairs, each for itself; the binding energy is proportional to the number of pairs, $n/2$. The latter forces produce a collective deformation, an effect of "collaboration," whose binding energy is proportional to n^2. Hence the pairing forces are predominant for smaller n's, and the deforming quadrupole forces for larger values of n.

The spectra of clouds with even n accordingly have the following properties. For smaller n's the cloud is spherical and we find vibrations around the spherical equilibrium shape. The quanta of these quadrupole vibrations jump in units of two and give as the first excited state a 2^+ state; at about twice the energy, the three states 0^+, 2^+, and 4^+ are usually found. Such spectra are frequently observed.

If n is large, the cloud is deformed and the nucleus becomes nonspherical. Most of the deformation resides in the cloud, though the nonuniform forces of the cloud upon the core also deform the core to a slight extent. The nucleus then possesses a large quadrupole moment. The numerical value turns out to be composed of about equal contributions from cloud and core deformation; the core, however, having so much more charge, is actually much less deformed. The nonspherical nucleus can perform rotations which give rise to the characteristic rotation spectra.

Figure 4 shows examples of cloud spectra of both types. Because of the collective nature of these vibrations and rotations, the excitation energies are considerably lower than the one-particle excitation energies. They depend critically upon the number of n of particles in the cloud. The higher n is,[1] the smaller are the excitations. When the cloud is nearly a closed shell, the excitations become rather high and similar to single-particle spectra.

[1] n is counted such that it is never larger than half the occupation of a shell. If a shell is more than half occupied, n is the number of holes.

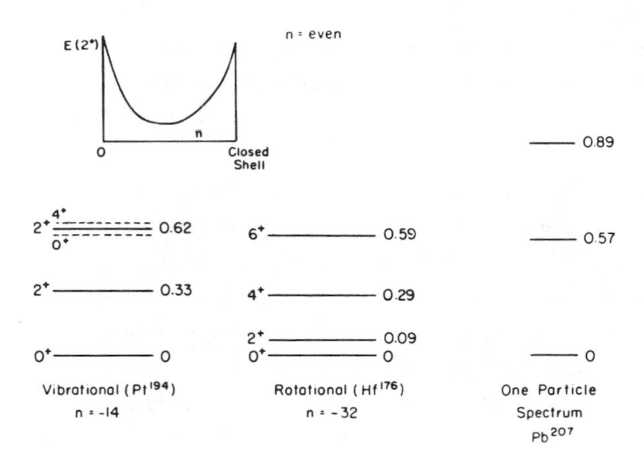

Figure 4.
Spectrum of cloud with sufficient number of particles. Pt^{194} is an example of a vibrational spectrum; Hf^{176} is an example of a rotational spectrum. The graph illustrates dependence of the excitation of the first excited state of the cloud as a function of *n*. At right a one-particle spectrum is added for comparison. (In the vibrational spectrum of Pt^{194} the two levels indicated by broken lines were actually not found. However, similar groups were observed with other examples.)

Clouds with odd *n* have a characteristic spectrum, which is a combination of collective terms and the single-particle properties of the odd particle. If one particle is added to a deformed cloud, one gets the characteristic rotational spectrum of a single particle in a rotating potential well. If an odd particle is added to a spherical cloud of even *n*, the first excited states are not the one-particle states of the odd nucleon, as several authors have recently shown very clearly, but are instead the first excited state of the cloud combining with the spin of the odd particle in its ground state. The cloud spectrum is the lowest of all nuclear spectra and must be excited first. The measurements of Crut et al. show this fact in a most telling way in Fig. 5. Here we see that the odd particle with spin 3/2 added to Ni^{62} or Ni^{64} gives rise to excited quartets; the odd particle with spin 1/2 added to Fe^{56} gives excited doublets.

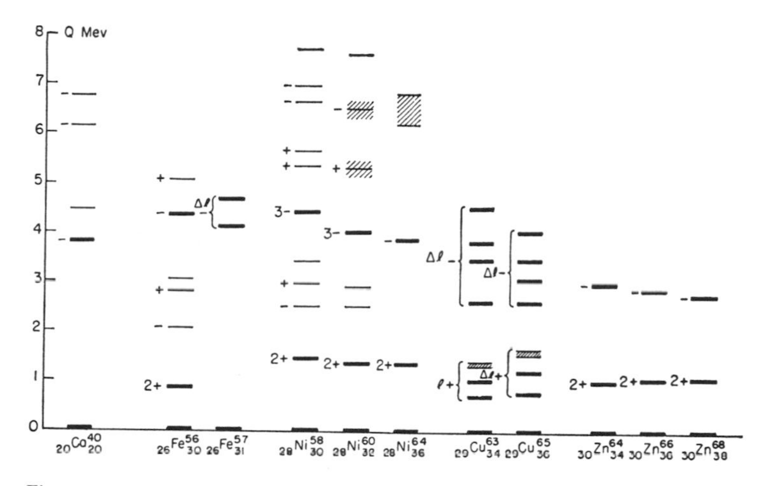

Figure 5.
Spectra near the elements of iron and nickel which illustrate the combination of single-particle states and cloud spectra. Measurements made by Crut et al. (private communication to N. S. Wall.)

So far we have discussed single-particle spectra and cloud spectra. For a third kind we look at the excitation of nuclei at much higher energy. It is known that all nuclei show a broad dipole resonance, the so-called giant resonance, at around 20 Mev—higher for light nuclei, lower for heavier ones. The excitation energy of the giant resonance is a smooth function of A and does not show any marked dependence on the shell structure. It is interpreted as a vibration of all protons against all neutrons in the whole nucleus and can therefore be considered as a dynamic motion of the core itself.

It is not the only way the core responds to outside influence. Recently B. Cohen [3] has found from inelastic proton-scattering experiments that all nuclei exhibit a 3^- resonance at roughly 3 Mev, a resonance which also was found with inelastic α scattering by several authors [4]. This resonance too does not show any dependence on closed shells and must be regarded therefore as a core octupole vibration. There exists probably a quadrupole

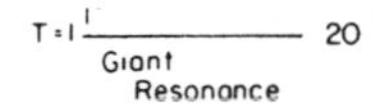

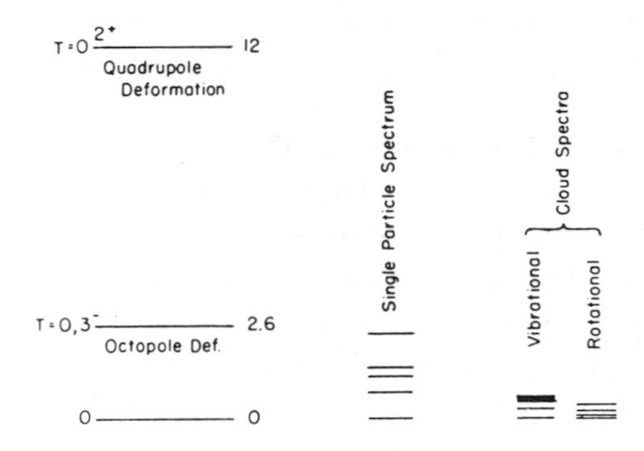

Figure 6.
The spectrum of the core. This is only a schematic presentation. For comparison, at the right a single-particle spectrum and the two kinds of cloud spectra are added.

vibration of the core which is not directly observed but indirectly inferred by Fallieros and others [5] from the fact that a neutron in its orbit around a closed shell can emit electric multipole radiation. This can only be if the neutron can produce a tidal wave in the core which acts as an accompanying charge. From the strength of this effect the quoted authors concluded that the core has a 2^+ vibration which lies for heavy nuclei at about 12 Mev.

Recently a number of physicists were able to excite these core vibrations with fast electrons [6]. The inelastic scattering of several 100-Mev electrons singles out high-energy collective core vibrations. According to Kendall et al., there are probably more vibrations present in the core. They find indications of a 4^+ excitation at 5 to 6 Mev. Figure 6 shows a typical core spectrum of a medium-heavy nucleus.

The excitations of a nucleus can thus be divided into three classes. The highest in energy are contained in the core spectrum. Some of them can be easily interpreted, e.g., the giant resonance as the motion of protons against neutrons; others, such as the octupole deformation, require still a better understanding as to their nature and energy. The relatively high energy is connected with the stability of closed shells. The second class of nuclear excitation contains the single-particle levels. They are much less energetic than the core excitations. Their character is well known from the numerous studies of the shell model. The third class contains the collective cloud excitations of vibrational or rotational character. They represent the lowest excitations, in particular for nuclei with many particles in open shells. The energies involved in these motions are low because of the following circumstances: The collective nature implies the simultaneous motion of many particles, which increases the inertia; in the core this is overcompensated by very strong restoring forces. In the cloud, however, there is no rigid structure which keeps the particles in well-defined states; the restoring forces against displacement therefore are very weak. Figure 6 gives a rough indication of the relative size of the three types of excitations.

Obviously the actual nuclear spectra are combinations of all three types. For example, any of the core excitations are widely broadened because of the many combinations between all types of excitations. The odd cloud spectra in Figure 5 are a combination of a single-particle term with a vibrational excitation.

The classification of nuclear excitation into three types should serve as a qualitative principle only. It is a theoretical model in the sense that it overstresses certain simplifying features which certainly exist, but which cannot be isolated as clearly as this description makes it appear. The excitation energies of the different types are not far enough apart to allow a clear distinction. This is true in particular for higher excited states. Hence the assignments cannot be as clear-cut as, for example, the assignment of molecular spectra as combinations of electronic excitation and atomic vibration and rotation.

We now turn to the wide field of nuclear reactions. Here also many models are in use to explain the different phenomena. We contend that again all useful models can be analyzed by starting with the independent-particle picture and introducing the nuclear interactions afterwards to modify and expand the model. The process taking place when a particle projectile enters a nucleus can be well described by similar models, such as the one used for the dynamics of nuclei. The nucleus is replaced by a potential well. Because of the probability of energy dissipation —it does not exist in nuclear ground states and only weakly in lowly excited states—we now add an imaginary part to the nuclear potential and obtain what is usually called the optical model. This model is extremely successful in describing the elastic scattering by nuclei and the "absorption" of nucleons, when we understand by the latter any process which is *not* elastic scattering. A good illustration of its success are the results of a determination of slow-neutron reactions by means of this model. It will not reproduce, of course, the compound-nucleus resonances, since they are produced after "absorption," but it reproduces excellently the gross-structure of the events—that is, the average values over neighboring resonances. If the energy is so low that only *S*-states are active, one finds resonances in the gross-structure whenever a standing wave can build up in the potential well. Figures 7 and 8 show the absorptive and the capacitive effect of these resonances, in the neutron strength function and in the potential scattering radius, respectively. The

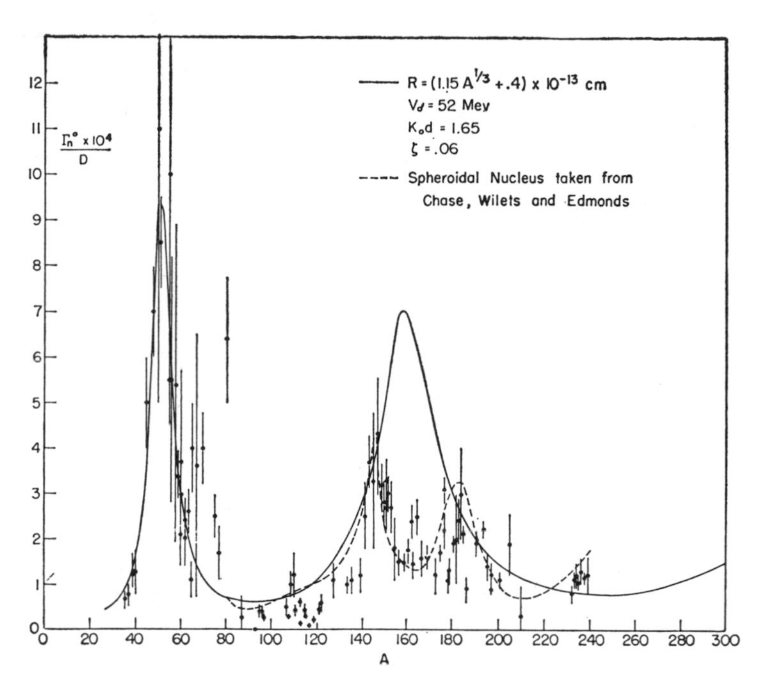

Figure 7.
Neutron strength function at zero energy as a function of the mass number. Theoretical curves correspond to optical model with a depth V_0, an imaginary part ζV_0, and a thickness of the surface d. The values of these constants are given in the picture. K_0 is the wave number of the entering neutron in the potential well. The solid theoretical curve corresponds to a spherical nucleus; broken theoretical curve corresponds to spherical nucleus according to Chase, Wilets, and Edmonds.

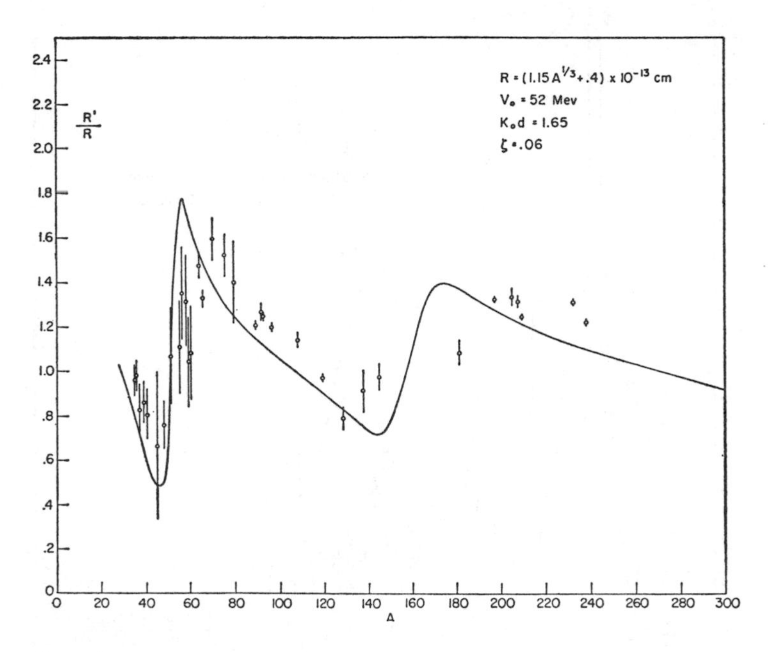

Figure 8.
Effective scattering radius at zero energy as function of mass number for same conditions as in Figure 7.

few deviations from the expected curves are now reasonably well understood and are traced to nuclear deformation (around the second resonance) and fluctuations in the imaginary potential because of the absence of levels of correct parity (around $A = 100$). The optical model is equally successful in the description of the scattering and absorption of nuclear particles at higher energy. The increase of the imaginary potential with energy and the superposition of more than one angular momentum prevent the resonances from being so spectacular as they are at low energy.

When one goes to the other extreme, high-energy and very strong absorption, as one finds it with 40-Mev α-particles, the

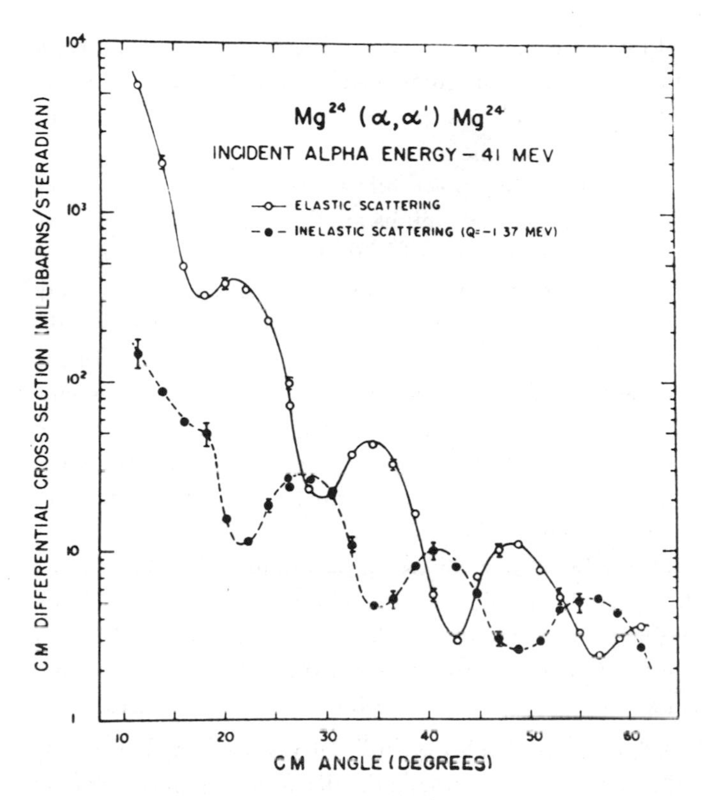

Figure 9.
Angular distribution of scattering of 40-Mev α particle on magnesium-24 as measured by Yavin and Farwell. We are here only interested in the upper curve which shows the elastic scattering. The experimental curve follows closely the expected diffraction pattern of a black sphere.

optical model would predict scattering patterns with maxima and minima in their angular distribution very similar to the optical scattering of black spheres. This prediction is borne out very well, as Fig. 9 indicates.

The optical model is a powerful tool for the analysis of nuclear reactions, but it does not tell us anything more about nuclear reactions or inelastic scattering than the total probability of such events. In order to find out more we must look more closely into the mechanism of what happens when a particle is "absorbed" by the nucleus, as understood in the language of the optical model. For this purpose we go back to the independent-particle model and study the modifications introduced by the interactive forces. What happens when a particle enters the nucleus and collides with one of the nuclear constituents? Figure 10 illustrates some of the possibilities.

(1) The incident particle loses a part of its energy by lifting a nuclear particle to a higher state. This will result in an inelastic scattering if the incident particle has enough energy left to get out again. One calls this process a direct inelastic scattering because it involves scattering with one constituent only.

(2) The incident particle transfers the energy to a collective motion, as is indicated symbolically in the second graph of Figure 10; this also is a "direct interaction."

(3) In the third graph the transferred energy is large enough to eject a nucleon from the target. This process will give rise to a direct nuclear reaction. In principle, it is not different from (1); it is the corresponding "exchange reaction."

(4) The incident particle can lose so much energy that it stays bound within the nucleus, and the transferred energy might be taken up by a low-lying nucleon so that it cannot leave the nucleus. We then obtain an excited nucleus from which no nucleon can escape. This state is bound to lead to further excitations of nucleons by internal collisions, in which the energy per excited particle in the average is further decreased, so that in

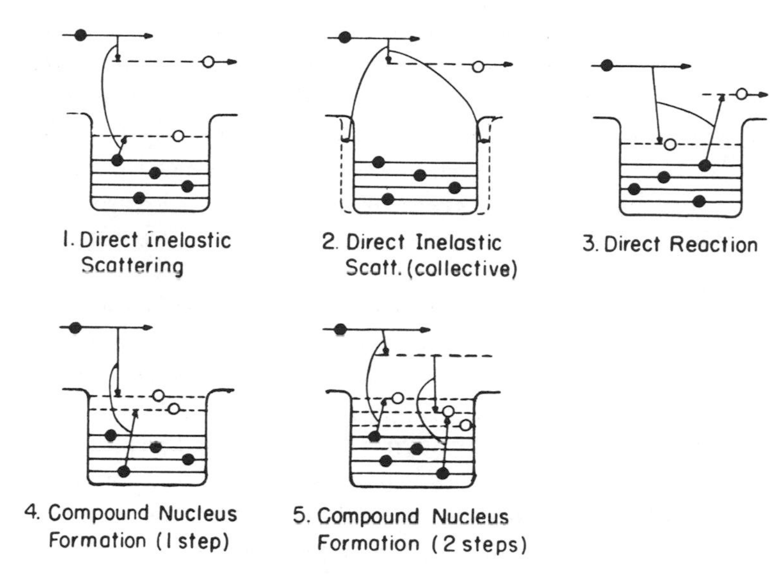

Figure 10.
Graphic representation of what happens when a particle enters a nucleus. The dotted deformation of the potential well in the second drawing should symbolize a collective vibration.

most cases no nucleon will be able to escape. Hence a state of very long lifetime is reached, and decay can take place only when one particle accidentally, by internal collisions, accumulates enough energy for a successful escape. This is what we call compound-nucleus formation. Energy might also be lost by radiation, after which particle escape is energetically impossible: The incident nucleon would have undergone a radiative capture.

(5) Compound-nucleus formation also can happen in two or more steps if after a process of type (1) or (2) the incident nucleon, on its way out, hits another nucleon and excites it in such a way that no escape is possible for either nucleon.

This scheme demonstrates how the assumption of relatively

weak interactions between nucleons in nuclear matter makes it possible to understand both the occurrence of direct reactions and the formation of a compound nucleus. The concept of the compound nucleus was first conceived by N. Bohr under the assumption of very strong interactions between nuclear constituents. In this case, the energy of an incident particle would be quickly and thoroughly distributed among all constituents. When the new ideas about weak coupling in nuclei became current, the compound nucleus picture was pushed into the background, and it was often asserted that all nuclear reactions should be direct ones. It is clear, however, that the model of weak interactions does not imply this at all. It implies that direct reactions will take place, but it does not exclude that, part of the time, the incident energy is transferred to a large number of constituents, in particular when the incident energy is not too high. In fact, once the compound state is reached, the lifetime of the state would be considerably longer than expected from a strong interaction model.

The series of events following the entry of a particle in a nucleus are symbolically described in Fig. 11. Here we see on the left the incident particle coming in and being scattered by the complex potential. Part of it is "absorbed," which means that it suffers a first collision, leading either to direct inelastic scattering or to direct reactions. A second collision still may lead to a reaction but with less probability because of the dissipation of the energy among three particles. From now on more and more collisions take place which rarely lead to an ejection of a particle but to a thorough interchange of energy within the nucleus. A compound state is formed. After many such interchanges, one particle (rarely the same as the incident) succeeds in escaping; the compound nucleus has decayed. It is seen how the weak interaction concept leads to the two-way street of direct reactions and compound-nucleus formation. It is also clear from

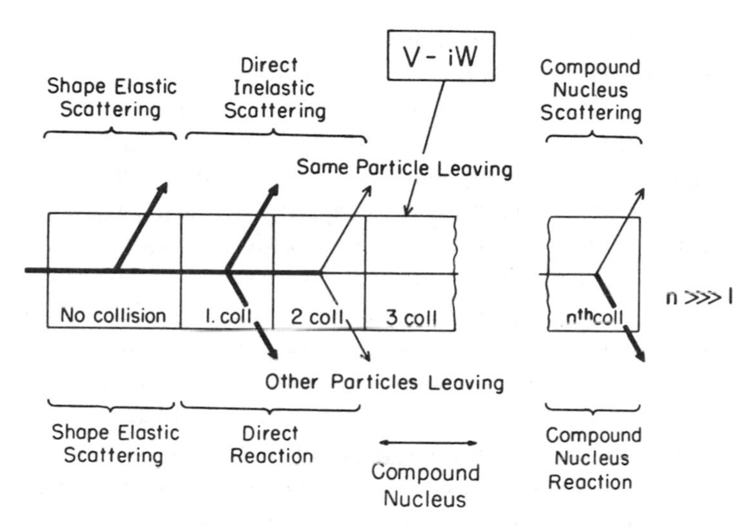

Figure 11.
Graphic representation of the course of a nucleus reaction. As long as no collision takes place, only shape-elastic scattering is possible. After a first collision direct reactions take place; later on, after many collisions, compound nucleus formation occurs.

this picture that the relative probability of forming a compound state will decrease with increasing incident energy. If the latter energy is very high, say larger than 30 or 40 Mev per incident nucleon, we expect that the first collision will almost always lead to a direct process. In these cases, a compound state (a state in which the excitation energy is distributed among many nucleons so that each of them is unable to escape unless one acquires an abnormally large amount) is going to be formed only after one or more nucleons have been ejected.

The probabilities of direct interactions can be calculated by well-known quantum-mechanical methods dealing with the collision of two particles in the nuclear potential well. Many important experimental facts about these reactions could be explained that way, in particular the angular distributions of reaction products, which are so different from what one expected in an evaporation model. The decay of the compound nucleus gives rise to the characteristic energy spectrum which is close to a Maxwell distribution of a given temperature and, of course, to an angular distribution symmetric about 90° and almost isotropic. These typical properties made it possible to distinguish between direct and compound reactions in a fairly reliable way.

Today we can study both types of reactions; Figs. 12, 13, and 14 show characteristic examples of this. Figure 12 presents a beautiful evaporation spectrum of inelastically scattered neutrons by copper with an incident energy of 7 Mev, as observed by D. B. Thomson in Los Alamos. There is no direct reaction visible because of the low incident energy. Figure 13 presents the spectrum of secondary protons emerging from an (*n-p*) reac-

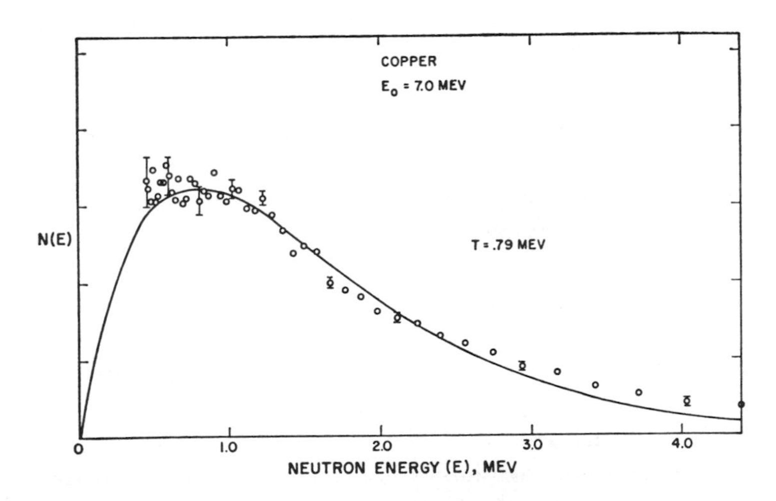

Figure 12.
Energy distribution of the outgoing neutrons $N(E)\,dE$ as a function of the energy E of the outgoing neutrons for the (n,n') reaction on copper with an incident energy of 7 Mev, according to measurements by David B. Thompson, Los Alamos, 1960.

tion with 14-Mev incident neutrons, as obtained by Facchini and collaborators in Milan; Figure 14 shows the secondary α-particles from an inelastic α scattering measured by Lassen and Poulsen in Copenhagen. In both cases we see the characteristic low-energy continuous-evaporation spectrum and, at high energies, a few well-separated peaks which correspond to the direct interactions. It is seen clearly that the direct interactions are preferably directed forward.

It is worth mentioning that, in principle at least, the direct processes and the compound-nucleus reactions could be kept apart by a pulsed-beam experiment. As shown in Figure 11, the direct processes take place immediately after the collision, within time intervals corresponding to the transit time of the incoming particle, which are of the order of 10^{-20} sec. The compound reactions, however, must wait for the decay of the compound nucleus, a time interval which is difficult to estimate, but which must be of the order of 10^{-16} or 10^{-17} sec. Evidently the times involved are much too short to contemplate an actual experiment at the present time.

Let me end this review with a slightly philosophical remark. Nuclear physics has shown the great variety of phenomena which one expects from quantum-mechanical systems of as many as a few hundred nucleons. Many unexpected and unusual shapes and motions were found and were often also understood. Let us compare this newly discovered world with the world of atoms and molecules. In many respects the situation is analogous: In both worlds we are dealing with quantum-mechanical systems of many particles. There are important differences, however. The fact that the stable units in the atomic world are uncharged opens up the possibility of composite units such as molecules, macromolecules, and crystals; it allows for the infinite variety of substances, living or dead, with which we are surrounded. No such nuclear aggregates can be formed, and hence the world of nuclei is somewhat poorer than the atomic world.

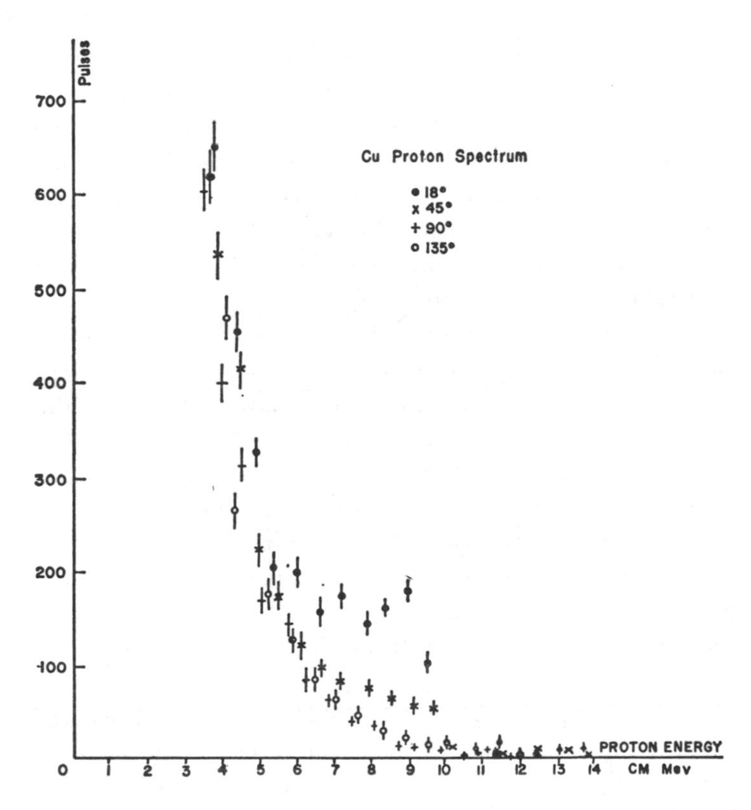

Figure 13.
Energy distribution of protons from the $n-p$ reaction with 14-Mev neutrons at different angles as measured by L. Colli, U. Facchini, I. Iori, G. Marcazzan, A. Sona (Milano, Italy). It is seen that the direct reactions are mainly visible in the forward direction.

Figure 14.
Energy distribution of inelastic scattered α particles from the (α, α') reaction on copper-63 and copper-65 (bottom curve) at different angles, according to measurements by N. O. Lassen and N. O. Roy Poulsen, Copenhagen. It is seen that the direct interactions are visible mainly at small angles.

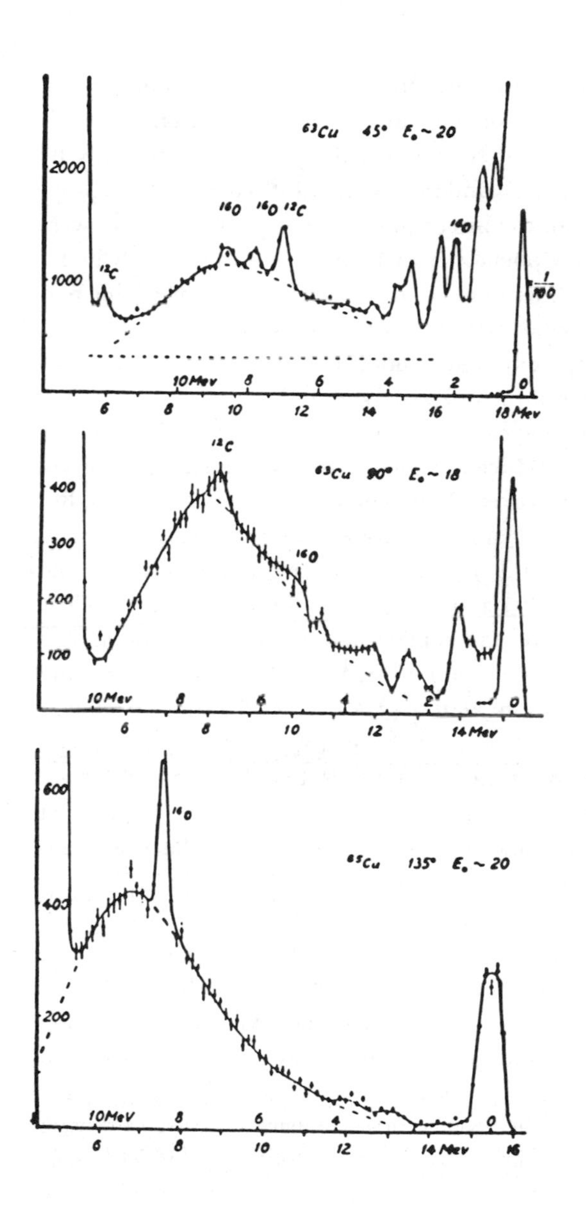

^{63}Cu 45° $E_0 \sim 20$
2000
1000
^{16}O ^{16}O ^{12}C
^{12}C
^{16}O
1/100
10 Mev 8 6 4 2 0
6 8 10 12 14 16 18 Mev
^{12}C
^{63}Cu 90° $E_0 \sim 18$
400
300
200
100
^{16}O
10 Mev 8 6 4 2 0
6 8 10 12 14 Mev
600
^{16}O
^{65}Cu 135° $E_0 \sim 20$
400
200
10 MeV 8 6 4 0
6 8 10 12 14 Mev 16

There is another difference from our human point of view. In our immediate environment atomic nuclei exist only in their ground state; they affect the world in which we live only by their charge and mass and not by their intricate dynamic properties. In fact, all the interesting nuclear phenomena we were talking about are phenomena which come into play only under conditions which we have created ourselves in accelerating machines. It is to some extent a man-made world.

It is not completely man made, however. The centers of all stars are regions of the universe where nuclear reactions go on, and thus where nuclear dynamics plays an essential role in the course of nature. Hence the nuclear phenomena are the basis of our energy supply on earth, in reactors as well as in the sun. But nuclear physics is even more important for the world in which we live from the point of view of the history of the universe. The composition of matter as we see it today is the product of nuclear reactions which have taken place a long time ago in the stars or in star explosions, where conditions prevailed which we simulate in a very microscopic way within our accelerating machines. Hence the material basis of the world in which we live is a product of the laws of nuclear physics. I cannot better illustrate the interconnection of all facts of nature, the tightly woven net of the laws of physics, than by pointing to the chart of abundances of elements in our part of the universe (Fig. 15). Each maximum and minimum in the curve of abundances corresponds to some trait of nuclear dynamics, here a closed shell, there a strong neutron cross section or a low binding energy. If the 7.65-Mev resonance in carbon did not exist, then, according to Hoyle and Salpeter, practically no carbon would have been formed and we would probably not have evolved to contemplate these problems. Whenever we probe nature—be it by studying the structure of nuclei, or by learning about macromolecules, or about elementary particles, or about the structure of solids—we always get at some essential part of this great universe.

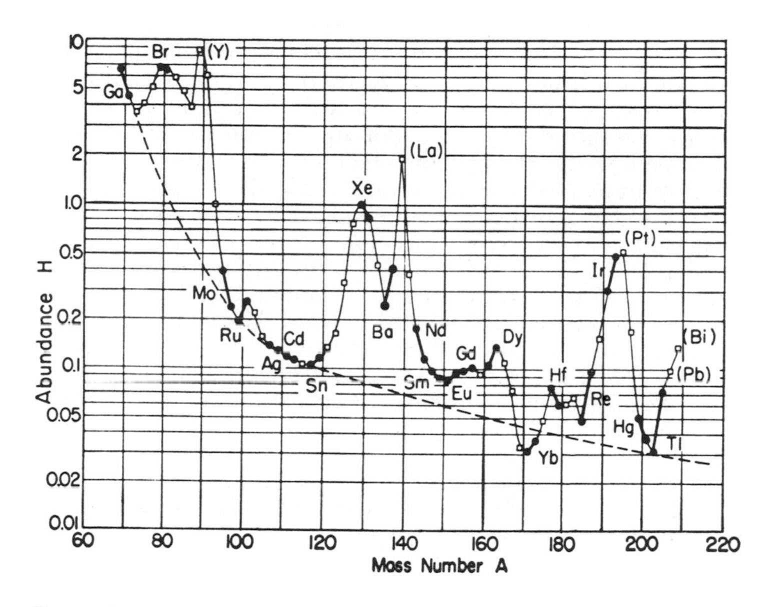

Figure 15.
Chart of abundances of elements in the universe according to H. E. Suess and H. C. Urey, *Revs. Mod. Phys.* **28**, 53, 1958.

References

[1] Derrick, Mustard, and Blatt, *Phys. Rev. Letters* **8,** 69, 1961.
[2] E. E. Salpeter, *Annals of Physics* **11,** 393, 1960.
[3] B. Cohen, *Phys. Rev.* **105,** 1549, 1957.
[4] J. S. Blair, G. W. Farwell, and D. K. McDaniels, *Nucl. Phys.* **17,** 641, 1960; M. Crut and N. S. Wall, *Phys. Rev. Letters* **3,** 520, 1960.
[5] Fallieros and Farrell, *Phys. Rev. Letters* **116,** 1960.
[6] H. Cranwell, R. Helm, H. Kendall, J. Oeser, and M. Yearian, to be published.

Quantum Theory and Elementary Particles

> All these things being considered, it seems probable to me that God in the beginning formed Matter in solid, massy, hard, impenetrable, moveable Particles, of such Sizes and Figures, and with such other Properties, and in such Proportion to Space, as most conduced to the End for which he formed them; and that these primitive Particles being Solids, are incomparably harder than any porous Bodies compounded of them; even so very hard, as never to wear or break in pieces; no ordinary Power being able to divide what God himself made in the first Creation. While the Particles continue entire, they may compose Bodies of one and the same Nature and Texture in all Ages: But should they wear away, or break in pieces, the Nature of Things depending on them would be changed. Water and Earth, composed of old worn Particles and Fragments of Particles, would not be of the same Nature and Texture now, with Water and Earth composed of entire Particles in the Beginning. And therefore, that Nature may be lasting, the Changes of corporeal things are to be placed only in the various Separations and new Associations and Motions of these permanent Particles.
>
> Newton [1]

Quantum Mechanics and "Permanent Particles"

In this well-known and justly famous statement, Newton recognizes the logical necessity of elementary particles in order to explain the existence of materials with well-defined properties, such as "Water" or "Earth," metal or mineral, liquid or gas, and with characteristic and ever-recurring qualities. Matter must be composed of some entities on which those qualities are based, entities that today are called atoms or molecules. Newton faces a problem, however: The elementary constituents of matter must possess specific properties that do not change with time; they should not "wear off in use," they should be immune to any rough treatment. He solves this problem by assuming that they are "incomparably hard" and indestructible by any ordinary power. But today we know that this is not so. Atoms can be

Synthesis of two articles originally in *Science* **149,** 1181, 1965; *Sci. Amer.* **218,** 5, 15, 1968.

broken by a very ordinary power, for example by lighting a match, but they still possess intrinsic qualities. They regenerate themselves whenever the original conditions are re-established. What Newton ascribes to the "first Creation" happens everywhere and at any time. We observe well-defined shapes and qualities without permanence of the unit itself.

Today we know what Newton did not know, that this is based upon quantum mechanics. The fundamental ideas of quantum mechanics can be understood in terms of the particle-wave duality. The motion of a particle corresponds in some ways to the motion of a wave. The electrons are bound to the positively charged atomic nucleus by electric attraction and, therefore, are confined by this force to a finite region around the nucleus. Any wave motion that is confined in space develops a series of well-defined standing waves whose characteristic patterns and frequencies are determined by the geometric properties of the confinement. The vibrations with the lowest frequencies have the simplest shape. Quantum mechanics connects a frequency ω with an energy E by the Planck relation $E = h\omega$; hence the state of lowest energy corresponds to the characteristic vibration of the electron wave in the confined region that has the lowest frequency. The properties of this vibration depend only upon the character of the confinement. Thus, whenever an electron returns to its lowest energy state in the atom, it assumes the same state with well-defined qualities. The permanence of atomic properties is explained not by the "hardness of the atoms," but by the nature of the confining field which by necessity produces always the same pattern of vibration.

These patterns reflect the symmetries of the confinement. The symmetries are important for the classification of the stationary states. For example, the symmetry of space determines the character and shape of the atomic states. It admits scalar waves with one component, spinor waves with two components, and so on.

The shape of the states follows from the spherical symmetry of the nuclear Coulomb field: The states of lowest energy must be simple spherical harmonics. (See Fig. 1 page 75.)

There is a second symmetry in atomic systems with more than one electron: The equality of the electrons gives rise to a symmetry in regard to the permutation of electrons. The permutation symmetry admits two alternatives: The quantum state may be symmetric or antisymmetric with respect to an exchange of particles. Nature chose the second alternative for electrons, which gives rise to the Pauli principle: Only one electron is allowed to occupy a given quantum state. We know today that it is a necessary consequence of the spinor character of electron waves. This is where the large variety of atomic shapes comes from, since electrons are forced into higher and different forms when the lower ones are occupied. Without this principle, all electrons in an atom would crowd into the lowest state and all atoms would have essentially the same shape, comparable to the noble-gas atoms. In many ways, the Pauli principle replaces the classical concept of impenetrability or hardness. Two identical particles, obeying the principle, can never be brought to the same place. It is therefore reasonable to reserve the term *particle* for the entities that obey the Pauli principle.

The spectrum of atomic energy levels reflects the basic symmetries. They cause a characteristic grouping of levels—the multiplets—whose multiplicity, wave form, and other properties are determined by the symmetry. We arrive in that way at a classification scheme of atomic levels by means of the quantum numbers of spin and angular momentum.

It is important to keep in mind that these symmetries do not determine all properties of the quantum states. They determine the general shape and many other features, such as the structure of the level spectrum and details regarding transition probabilities. They do not give the size or the energies of the quantum states. These properties are determined by the strength and by the nature of the forces with which the particles are bound within the system. The symmetries alone are not sufficient for a complete description of a system; a knowledge of the dynamic conditions is required.

Let us now return to Newton's remarks. Does quantum mechanics of atomic structure fully remove the difficulty that Newton brought forward? It leads to an essential insight into the origin of fundamental shapes in nature: Intrinsic shape and ever-recurring properties of atoms are understood without assuming that the atoms are "incomparably hard." But Newton would not have been completely satisfied with this answer, because our conclusions are based upon the existence of other particles, electrons and nuclei, which themselves possess intrinsic properties, such as mass, charge, spin, and magnetic moment. So the question obviously is raised again on a new level. Are the atomic constituents incomparably hard? Is there an ordinary power that can take them apart?

As far as the nuclei are concerned, the answer is known. Nuclei can be taken apart by ordinary power; they consist of protons and neutrons. A strong attractive force keeps these particles together. The nature and origin of this nuclear force is not understood yet, but we are well acquainted with its general properties. The sizes and the characteristic binding energies of nuclei are determined by this force, and it is of interest to compare atomic with nuclear sizes and energies. The atom is held together by an electric force whose potential is given by $-e^2/r$ with $e^2/hc = 1/137$, e being the electronic charge, r the distance from the center of the field, h Planck's constant, and c the velocity of light. From this it follows that atomic sizes are of the order of a Bohr radius $a = h^2/me^2$ and atomic energies of the order of the Rydberg $\mathrm{Ry} = me^4/h^2$, with m being the electron mass. The nucleus is held together by a nuclear force whose potential is somewhat more complicated, but whose most relevant contribution has the Yukawa form

$$-(g^2/r)\,[\exp(-r/r_0)]$$

where $g^2/hc = 0.08$ and r_0 is the range of nuclear forces. If, for a moment, we set the exponential factor equal to unity, we get the same kind of potential as in atoms and would expect the size of nuclei to be of the order of a "nuclear Bohr radius" $a_N = h^2/Mg^2 = 2.8 \times 10^{-13}$ cm where M is the nucleon mass, and nuclear energies to be of the order of a "nuclear Rydberg" $\mathrm{Ry}_N = Mg^4/h^2 = 5$ Mev. These values do indeed give a good orientation as to the sizes and energies of nuclear phenomena. The fact that r_0 is of the same order as the range of nuclear forces is a justification for the omission of the exponential factor in the Yukawa force.

The intrinsic shapes and forms of nuclei are determined by the same symmetries as the atomic shapes. This is why nuclear physics is similar in so many ways to atomic physics. There is one obvious difference between nuclear and atomic dynamics. In the atom the nucleus provides a massive center with an electric charge ranging from one to more than 100 units in the heaviest man-made elements. Furthermore, the electric force is not limited in range. Hence the attractive force of the nucleus is the overriding force acting on each electron. The dynamics within the nucleus is more "democratic": There is no central particle with an overriding attraction. All constituents have nearly the same mass and attract each other by about the same nuclear

Figure 1. (See opposite page.)
Similar periodicities in atoms and nuclei are evident in plots of (a) atomic ionization energies compared with (b) nuclear excitation energies. Largest energies occur for inert gases and magic nuclei, both of which correspond to closed shells. Nuclear excitations are shown for first excited levels in even-even nuclei; the solid lines connect all nuclei that have the same proton number.

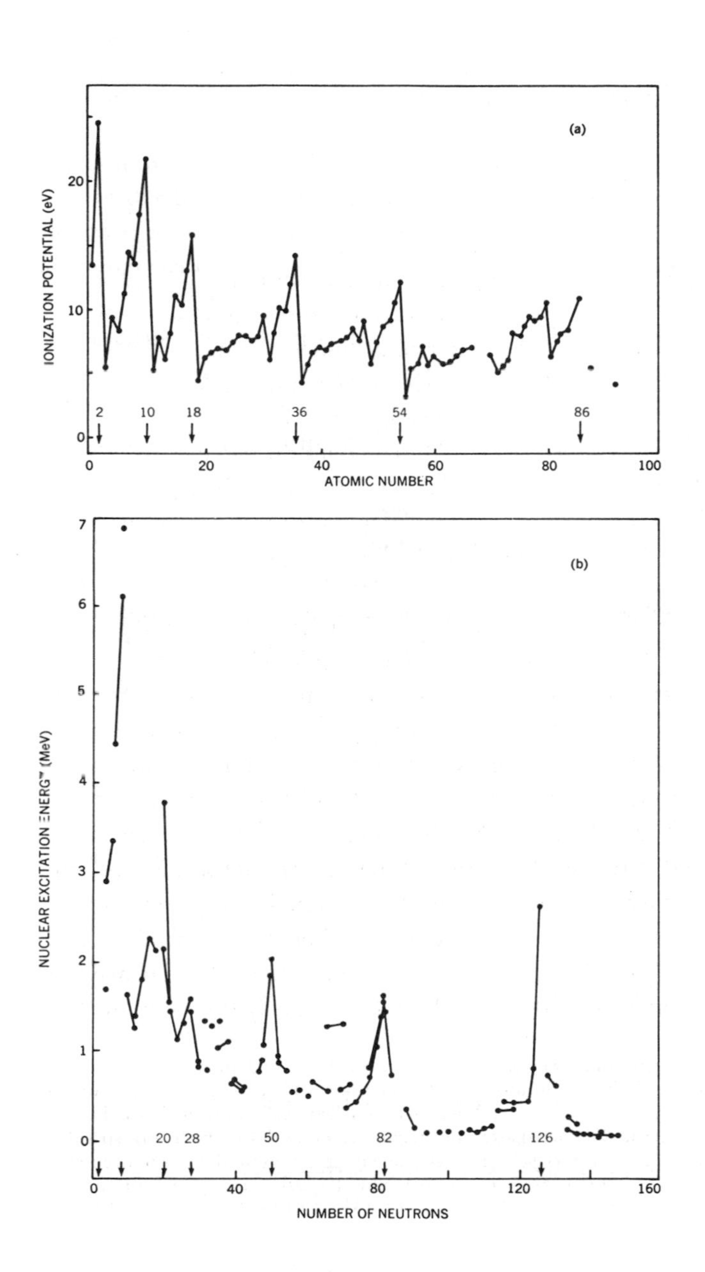

(a)
IONIZATION POTENTIAL (eV)
0
10
20
2
10
18
36
54
86
0
20
40
60
80
100
ATOMIC NUMBER
(b)
NUCLEAR EXCITATION ENERGY (MeV)
0
1
2
3
4
5
6
7
20
28
50
82
126
0
40
80
120
160
NUMBER OF NEUTRONS

force. Each particle moves and produces its vibrational pattern in the combined field of force of all other particles.

Nevertheless, there are many similarities between atomic and nuclear spectra. Again we find the same typical vibrational patterns and multiplets. Because the Pauli exclusion principle also applies inside the nucleus, no two equal particles can be in the same state.

Another striking similarity is found in the shell structure of both atoms and nuclei. Whenever the number of constituents—electrons in the first case, protons or neutrons in the second—reaches certain values, closed shells of increased stability are formed. Accordingly there is a periodic table not only of elements but also of nuclei (Figure 1).

A new symmetry appears in nuclear spectra that does not exist in atomic spectra. Since the nuclear force does not distinguish a proton from a neutron, the nuclear states that differ only in the replacement of a neutron by a proton must have the same structure and therefore the same energy. This symmetry is not quite exact because the replacement changes the number of units of electric energy. The electric energies, however, are smaller than the nuclear energies; also they can be easily calculated and subtracted. This symmetry is called isospin symmetry for reasons that need not concern us here. It gives rise to groups of states called isospin multiplets.

When a nucleus is raised to an excited state and falls back to a lower energy state, it can get rid of its energy not only by emitting a light quantum, as an excited atom would do, but also—when the excitation is high enough—by emitting a lepton pair: the simultaneous emission of an electron (either a negative electron or a positron) and a neutrino. Both particles are created at a moment of transition from the higher to the lower nuclear state, just as the light quantum is born when a similar transition occurs. This new emission is called the beta decay of the nucleus.

Because nature always creates particles and antiparticles in pairs, the companion particle created with the electron is understood to be an antineutrino and the one created with the positron to be a neutrino.

This emission has several unusual features. One is that electric charge is emitted; hence the nucleus must change its overall charge in the opposite direction when the emission takes place, thereby ensuring the conservation of charge. What happens is that one of the nuclear protons changes into a neutron if a positron-neutrino pair is emitted, and one of the neutrons changes into a proton when an electron-antineutrino pair is emitted. Accordingly the charge of the nucleus is no longer an absolutely fixed quantity; it can be changed by beta decay. What cannot change in any transition is the total number of protons and neutrons, which is called the nucleon number. Nuclei that have the same number of nucleons cannot really be considered as different systems even if they have a different number of protons. They are nothing but different quantum states of the same nucleus.

The Baryon Spectrum

Would Newton have been satisfied at this point? Not completely; the number of elementary particles is essentially reduced to three: proton, neutron, electron. (We do not count the light quantum among particles, since it is the quantum of the electromagnetic field and obeys Bose statistics. The neutrino is excluded because it never appears as a constituent of matter.) But the existence of these particles still remains an assumption: They have "God-given" properties and may have to be "incomparably hard" so as not to change their properties when in use.

Let us first look at the situation regarding the proton and the neutron. So far nobody has taken a nucleon apart. The Rutherford of this stage is not yet known. It seems probable, however,

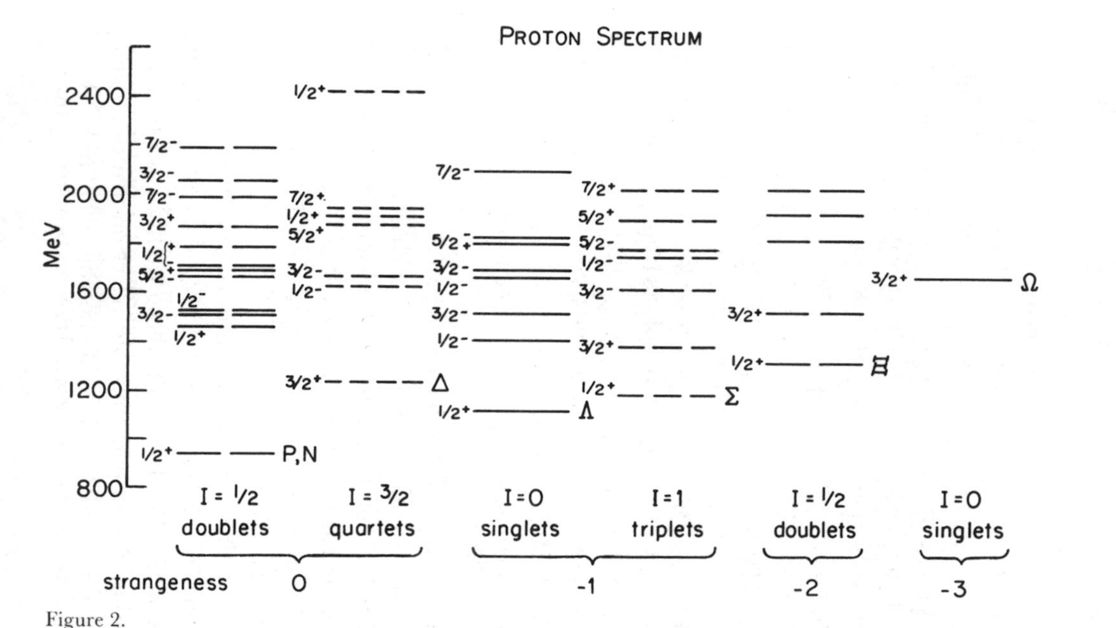

Figure 2.
The baryon spectrum is composed of the nucleon (P, N) and its various excited states. The states are arranged in columns according to their multiplicity and strangeness. The letter I denotes isotopic spin; the multiplicity is given by $2I+1$. In the subnuclear spectrum of the proton the ground state is taken to be the mass energy of the proton, 0.938 Gev. The number to the left of each state indicates spin angular momentum and parity (+ or −). The symbol to the right is the name of the state.

that the nucleon is not "incomparably hard" either. Indications of an internal structure are clearly present; there exists a spectrum of excited states of the nucleon (Figure 2). They are not usually called excited states, but the observed phenomena can hardly be interpreted differently. What do we observe? When the nucleon is exposed to any kind of high-energy beams, it is transformed into short-lived states of higher energy, which are known under various names, such as "hyperons" or "resonances." The name "baryon" will be used for that which appears in the form of proton or neutron, or in the form of its excited states. In the spectrum of baryon states, the proton and the neutron figure as the ground states—they form a ground-state doublet. Precisely speaking, the neutron also is an excited state of the proton. All other states can be reached by supplying the nucleon in the ground state with the necessary excitation energy in one form or another. Some of the excited states have different charges from the ground state; some have different "strangeness" or hypercharge—a new property which turns up for the first time in these phenomena. The excited states return to the ground state in one or several steps, with the emission of π-mesons, K-mesons, light quanta, or electron-neutrino (lepton) pairs.

The charged mesons and the lepton pairs are charge carriers and therefore are emitted when there is a charge difference between excited and ground state; the K-mesons are also strangeness carriers and are emitted when the strangeness changes. The K-meson carries a positive unit of strangeness, the anti-K-meson ($\bar{K}$) carries a negative unit.

In atoms, transitions between quantum states take place mostly by light emission or absorption, that is, by coupling with the electromagnetic field. In atomic nuclei we find, in addition to light emission, lepton-pair emissions (electron-neutrino pairs), which are produced by weak interaction coupling with the

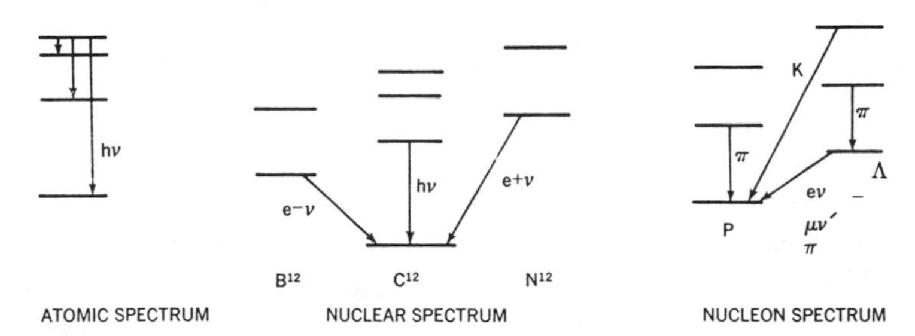

Figure 3.
Field quanta emitted in three types of spectra. Light quanta (electric interaction), lepton pairs (weak interaction) and mesons (strong interaction) characterize atomic, nuclear, and nucleon transitions, respectively.

lepton field. In the baryon spectrum we find, in addition to those two kinds of emissions, transitions with meson emission, which is transacted by the strong interaction of nucleons with a meson field (Figure 3). All three couplings are active in any of the three cases. But the energy differences between atomic states are too small to allow the emission of lepton pairs, for which an energy of at least 0.51 Mev is needed because one of the leptons must be an electron; the differences between nuclear states are large enough for lepton-pair emission (β-radioactivity) but too small for the emission of mesons, the smallest of which has a mass of about 140 Mev. In the spectrum of the baryon, however, the energy for transitions can be paid with the large units of a new currency—the mesons—although the ordinary currencies, light quanta and lepton pairs, are not excluded.

Let us quote a few examples of transitions between nucleon states: The simplest example is the emission of a π-meson in the transition from the first excited baryon state, a multiplet with the isotopic and ordinary spin of 3/2. This state has the same strangeness as the ground state; the transition is therefore accompanied by the emission of a π-meson. The charge of the emitted π-meson depends on the charge difference between the two combining

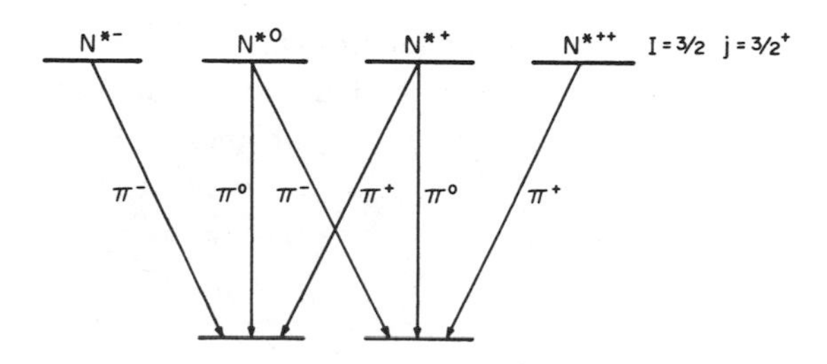

Figure 4.
Decay of delta state of the baryon is accompanied by emission of photons or pions. The emitted quanta carry off nearly 300 Mev of energy, allowing the delta baryon to fall to the nucleon ground state. The delta state is an isotopic quartet with four substates: one negative, one neutral, one singly positive, and one doubly positive. In falling to the ground state the first and fourth substates must change by one unit of charge, which is accomplished by the emission of a charged pion. The second and third substates need not change sign.

states (see Figure 4). Another example would be the transition from a highly excited state of strangeness different from that of the ground state: There a K-meson would be emitted, in order to carry away the difference in strangeness. An odd situation occurs with the lower excited states of different strangeness such as the ones designated by the symbols Λ, Σ, Ξ, Ω. They cannot de-excite by K-emission into the ground state, because the mass of the K-meson is higher than the energy difference. These states, therefore, would be stable if the conservation of strangeness were an exact law (as is the conservation of ordinary charge). In fact, however, strangeness is conserved in all interactions except the weak interaction. Therefore, there exist very slow transitions from those states to the ground state with emission of π-mesons or lepton pairs, mediated by weak interactions. Hence, the lowest states with strangeness different from zero are metastable and decay slowly into the only real stable state, which is the proton.

The excitation of these metastable states takes place mostly in a two-step process: First the nucleon is excited into one of the higher states, without change of strangeness, by proton collision or pion absorption; then a transition to a state of different hypercharge takes place, with the emission of a K-meson. This is called *associated production,* since the end product consists of two entities of opposite strangeness: an excited baryon and a K-meson.

The Meson Spectrum

Experiments with high-energy accelerators have revealed not only a spectrum of excited states of the nucleon but also a second spectrum: the spectrum of mesons, or boson spectrum (Figure 5). A careful analysis of the mesons produced by high-energy collisions has revealed that the π-meson and the K-meson are not the only forms in which mesons appear. There exists a series of excited states referred to by various letters: ρ-meson, ω-meson, η-meson, and so on, of which the π- and the K-meson are the lowest states. In fact, neither the π- nor the K-meson itself is really stable. Both decay by weak interactions into leptons. Hence they should be considered as true ground states of the meson spectrum only if the weak interactions are neglected.

In transitions from an excited meson state to a lower one, the energy difference also is emitted mostly in forms of mesons. For example, the so-called ρ-meson decays into two π-mesons. We can interpret this as a transition from the ρ-meson state to the lower π-meson state, with the emission of another π-meson. Figure 5 shows the most important meson quantum states known today and indicates their quantum numbers. Here, as in the baryon spectrum, we find the same quantum numbers as in nuclear spectra—ordinary and isotopic spin and parity—and also the new strangeness quantum number.

The existence of such excited meson states is perhaps not so surprising as one might think. Let us consider the situation from the point of view of the analogy between light quanta and

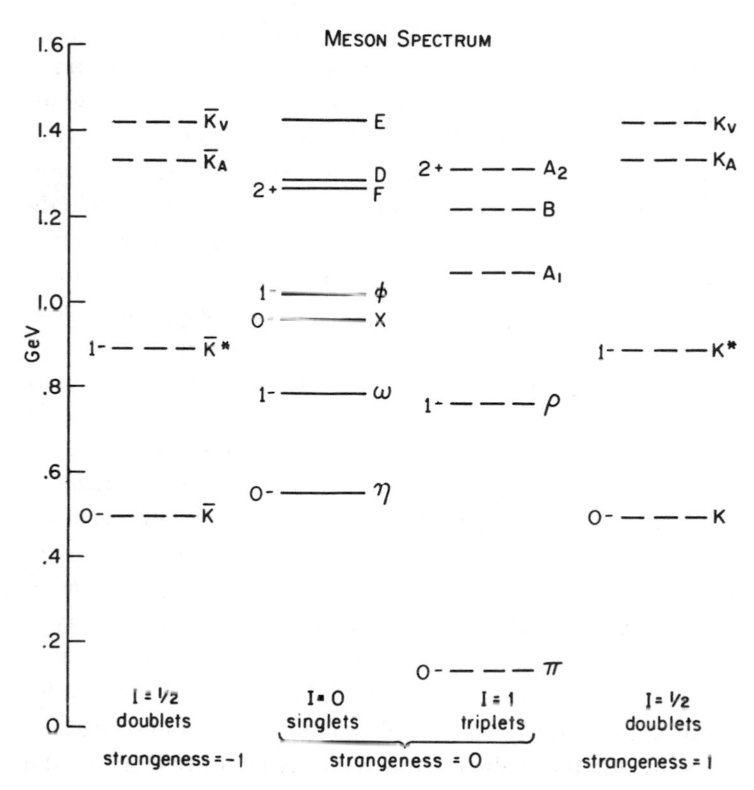

Figure 5.
Meson spectrum, like the baryon spectrum, can be arranged so that the various states fall in columns according to their multiplicity and strangeness.

mesons. Both entities are quanta of a field; the coupling of the quantum of the electromagnetic field with its source (the charge) is determined by the small constant $e^2/hc = 1/137$; it is a weak coupling. The coupling of the nuclear field to its source (the nucleons) is very much stronger. The corresponding magnitude G^2/hc is about 15. This is much larger than the magnitude $g^2/hc = 0.08$ that was used in estimating the strength of nuclear forces within nuclei; the nuclear forces have the peculiar property of being quite weak between nucleons whose relative momentum is nonrelativistic, as it is in the case of motion within nuclei. For that special situation the relevant coupling is reduced by a factor $g^2/G^2 = (m_\pi/2M)^2$ where m_π is the pion mass and M is the nucleon mass. It has its large value, however, under general conditions such as those for fields acting between particles of high relative momentum or between particles and antiparticles.

It is because of this circumstance that we can have a theory of nuclear structure based upon relatively weakly interacting proton-neutron systems without recurrence to the higher baryon states. If the relevant interaction constant in nuclear structure were as large as G, the nuclear excitations would be of the order of the baryon excitations; nuclear physics and elementary particle physics would be as closely related as meson physics and baryon physics.

The Consequences of Strong Couplings

A very strong coupling between field and source has a number of consequences, some of which can be understood by extrapolation from electrodynamics. It is known, for example, that two light quanta interact weakly with each other. If the coupling constant were larger than unity, however, the interaction would become large and would be comparable with the energy of the quanta. It would not be surprising, then, to find states in which several field quanta are bundled together. Such bundles are perhaps an appropriate description of the nature of excited meson states. There remains a question why no meson with rest mass zero exists in analogy to the light quantum. Is this also a consequence of strong interaction, or is there an essential difference between electromagnetic and mesonic fields? This is a most interesting problem which at present is left in complete darkness.

The above-mentioned interaction of light quanta comes from the fact that the two quanta can form virtual electron-positron pairs. Since the coupling constant between nucleons and mesons is large, the virtual pair states would play a much more important role in meson states. In fact, it would not be unreasonable to consider the meson states as states of a baryon-antibaryon system. There cannot be an essential difference between a bundle of field quanta and a state of a baryon-antibaryon system, since the former can produce the latter and vice versa. Because of the strong interaction, any such bundle will contain a considerable fraction of baryon-antibaryon pairs of equal spin and symmetry.

Not only mesons but also baryons should really be considered as surrounded by virtual baryon-antibaryon pairs. After all, the strong meson field in the neighborhood of the baryon must also give rise to virtual pairs. The physical baryon and the physical meson are in fact extremely complicated systems that can be described as mixtures of many different states: They contain any number of baryon pairs and meson bundles, compatible with the quantum numbers. The basic difference between the baryon and the meson states lies in the total number of baryons present. The baryon spectrum contains all states of matter in which this number is unity (the antibaryons are counted negative); the meson spectrum contains the states with $B = 0$. The spectrum of states with $B > 1$ would contain the spectra of nuclei with a nucleon number $A = B$ and also the spectra of the isobaric hyper-nuclei.

Here we recognize the reason why, in the meson spectrum, even the lowest states are metastable and decay by weak interaction into lepton pairs, whereas the ground state of the baryon spectrum—the proton—is really stable, and so are the ground states of most spectra with $B > 1$. The baryon number B is a quantity that is conserved in all interactions; hence only mesons can disappear by disintegrating into leptons; baryons must remain forever, a guarantee for the stability of our world.

When looking in this way at the baryons and mesons, one might be prompted to say that a meson is nothing but a given state of a baryon-antibaryon system or that any given state of the baryon is nothing but a combination of another baryon state and a few mesons. In other words, any of these states could be thought to be a combination of others. An ambitious attempt to get a self-consistent description of a group of particles in this way is known under the name of "the bootstrap method"; one requires that the masses and interaction constants should be such that one obtains self-consistent results in any of the possible combinations.

The complex nature of the physical baryon or meson is used today for the explanation of interaction processes by picking out a particular feature of the mixed surrounding of the particles that may be essential for certain interactions. For example, a collision in which one unit of charge or strangeness is transferred from one particle to the other is then described as the exchange of one meson carrying these properties. Such semiquantitative methods succeed from time to time in explaining a few characteristic experimental features. One tries to single out a typical Feynman diagram that symbolizes one of the many possible interaction mechanisms, ascribing to it greater importance than to all other possible ones. This is known under the name of *peripheral processes*.

New Symmetries and the Quark Model

The two new spectra, the baryon spectrum and the boson spectrum, are not yet understood. They seem to indicate some internal structure of these entities, in the same sense as the atomic and the nuclear spectrum are reflections of the internal dynamics of the atom and the nucleus. In the latter cases, however, we know the dynamics; we know that the atomic spectra represent the quantum states of electron motion in the Coulomb field and that the nuclear spectra represent the quantum states of the nuclear motion under their mutual attraction. This knowledge allows us to establish the rules of symmetry that determine the relevant quantum numbers and the corresponding multiplets. It also allows us—in principle, at least—to calculate the energies, sizes, and other properties of the quantum states.

The situation is different with regard to the baryon and boson spectra. We have no definite idea yet as to their dynamic basis. We cannot, therefore, predict any symmetry rules for them. However, empirical inspection of the spectra definitely reveals multiplet structures, which indicate the validity or approximate validity of certain symmetries in a yet unknown dynamics. What are these symmetries? First of all, one is able to ascribe to each state a total angular momentum J leading to the usual $(2J+1)$-fold degeneracy. This is obviously a reflection of the rotational symmetry of the situation and a consequence of the validity of our fundamental quantum-mechanical concepts. The decay patterns of unstable configurations exhibit the same geometrical shapes in simple spherical harmonics—which are the typical form elements of this symmetry. In fact, one has more occasion here than in the study of atomic or nuclear spectra to determine the angular momentum directly by the geometrical shape. This is because in high-energy processes particles are produced more often in polarized states, that is, in states in which the angular momentum has a specific direction. The angular distribution of the decay products reflects the special symmetry of the quantum state.

It is less obvious that there are also isotopic-spin multiplets, the characteristic groups of levels of almost equal energy which differ only in electric charge. The ground-state doublet (neutron-proton) is one example; the three π-mesons, π^+, π^0, π^-, are another. It seems that the charge independence found in nuclear structure carries over into the baryon and boson structures. (Or, rather, the observed charge independence in the baryon and meson structures is the basic fact, from which it would follow that interactions between nucleons and hence nuclear structure are also independent of charge.) The most interesting observation, however, is the presence of an additional symmetry: It has to do with the hypercharge or strangeness. As mentioned before, each quantum state of the nucleon and of the boson can be ascribed a strangeness quantum number. Evidently, the strangeness quantum number plays an important part in the ordering of these states. Is there any way in which it reflects a new symmetry?

This new symmetry has recently been discovered and is known as $SU(3)$ *symmetry*. The $SU(3)$ group that underlies this new symmetry is a generalization of the $SU(2)$ group. The latter is the "special unitary group" based on two fundamental states and is well known as the group which underlies the ordinary spin or the isotopic spin formalism. It is based on the "dichotomy" of two fundamental states—for example, the proton-neutron pair—and analyzes the properties of nuclear systems, derived from the invariance to an exchange between that basic pair. It is well known that this formalism leads to a multiplet structure in the spectrum. We have, of course, the basic proton-neutron doublet (isotopic spin 1/2); furthermore, a system of two nucleons gives rise to singlets and triplets of isotopic spin 0 and 1, a system of three nucleons gives rise to doublets and quartets of isotopic spin 1/2 and 3/2, and so on.

The $SU(3)$ group starts with a "trichotomy" of three states. Two of these represent a basic isotopic doublet with zero strangeness; the third is an entity whose isotopic spin is zero, but it carries a unit of strangeness that is usually assumed to be negative. Obviously, in the first attempts in this direction, the proton, the neutron, and the hyperon were considered to be the basic states of this fundamental triplet. Soon it turned out that the situation is, in fact, more complex and more interesting. The basic states of the "trichotomy" can be identified with three types of a hypothetical subparticle: the quark. Let us call the three types p, n, and λ. The first two quarks are so designed because they form an isotopic doublet of isotopic spin $I = 1/2$ just as the proton and the neutron do. The λ-quark is an isotopic singlet of isotopic spin $I = 0$. All three quarks have an ordinary spin of 1/2 unit and are very much alike except for their electric charge. Here a curious assumption is introduced: The p-quark is assigned a charge that is $+2/3$ times the unit charge; the n-quark is assigned a charge of $-1/3$ unit. Such fractional charge units have never been observed in nature. The λ-quark is assumed to have a strangeness quantum number S of -1, whereas the n- and p-quarks have S equal to 0. The charge of the λ-quark is $-1/3$ unit. This set of three quarks is provided with a set of antiparticle quarks: $\bar{n}$, $\bar{p}$ and $\bar{\lambda}$.

The baryon is supposed to consist of three quarks. According to the quark model, baryon states with S equal to 0 are composed of three quarks of the n or p type. Three p-type quarks would give a charge of three times $+2/3$, which is $+2$. Three n-type quarks would give a charge of -1. With such combinations we can obtain charges -1, 0, $+1$, $+2$, which satisfy all the charges of baryons with S equal to 0. The proton, for example, would consist of one n-quark and two p-quarks, the neutron of two n and one p. The states with S equal to -1 consist of one λ-quark and two quarks of the n or p type. This gives rise to charges of -1, 0, and $+1$. The states with S equal to -2 and S to -3, respectively, have two or three λ-quarks, which results in charges of -1 or 0 for S equal to -2, and a charge of -1 for S equal to -3. These are just the charges found among baryon states with a strangeness of -1, -2, and -3.

Let us now look at the isotopic spins of the different baryon states. What kind of baryons can we build if each baryon must be a combination of three quarks? If all three are of the p or n type ($I = 1/2$, $S = 0$), we get combinations with $I = 1/2$ or $3/2$ but $S = 0$; if two are of the (p, n) type and one is of the λ type ($I = 0$, $S = -1$), we get combinations with $I = 0$ or 1 but $S = -1$; if one is of the (n, p) type and two of the λ type, we get combinations with $I = \frac{1}{2}$ but $S = -2$; if all three are λ type, we get $I = 0$ and $S = -3$. These are exactly the quantum number relations found in the baryon spectrum: We find isotopic doublets and quartets ($I = 1/2, 3/2$) with $S = 0$, singlets and triplets ($I = 0, 1$) with $S = -1$, doublets ($I = 1/2$) with $S = -2$, singlets ($I = 0$) with $S = -3$. In this tentative picture the quantum states of the baryon are different dynamical forms of a three-quark system; the combination that makes up the proton has the lowest energy.

The quark model also provides a scheme for building mesons. There is an important difference, however: Mesons are energy quanta that can be emitted and absorbed freely. This means that quarks could enter into their structure only as particle-antiparticle pairs. Such pairs can be considered a form of exchangeable energy currency because the members of a pair can annihilate each other and thus change into other forms of energy, or they can be made from other forms of energy by the process of particle-antiparticle creation. For this reason the quark model assumes that mesons consist of a quark and an antiquark bound together. The emission of a meson would be equivalent to the creation of a quark-antiquark pair. The different combinations of the three types of quark and antiquark give rise to most of the observed meson types (Figure 6 and 7).

Let us now look at the meson spectrum. Here we observe the following characteristics: (i) a symmetry with regard to positive and negative strangeness, any particle in one group having its antiparticle in the other; (ii) the fact that the $S = 0$ mesons are their own antiparticles; (iii) only $I = 1/2$ for $|S| = 1$. This points to the fact that the mesons behave as if they were combinations of one quark and one antiquark (Figure 8). If the two quarks are of the type ($I = 1/2$, $S = 0$), such a combination would yield entities with $I = 0$ or 1, and $S = 0$, and they would be their own antiparticles. If one quark is of the second type ($I = 0, S = -1$), one gets a meson of isotopic spin 1/2 with $S = \pm 1$, depending upon whether the quark or the antiquark is of the second type. The $S = +1$ and $S = -1$ combinations then are each other's antiparticles.

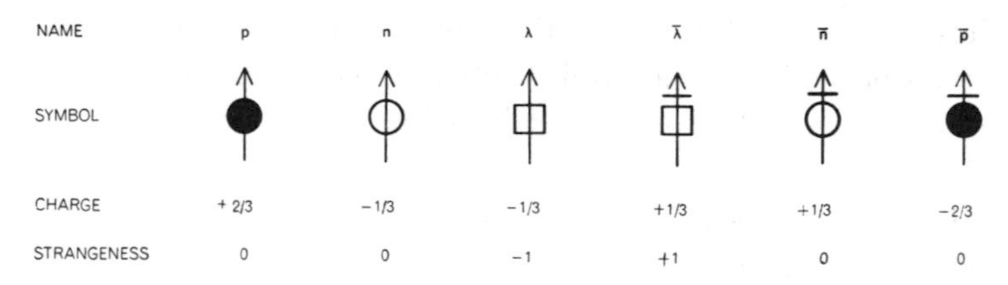

Figure 6.
Three quarks, and their antiparticles (right), have been proposed as the hypothetical building blocks of baryons and mesons. Unlike all particles discovered so far, quarks would carry less than a whole unit of the charge of the electron. The lambda (λ) quark is provided with a negative unit of strangeness. Each quark and antiquark carries half a unit of spin, the direction of which is shown by an arrow. As in the world of known particles, the hypothetical antiquarks mirror the properties of the quarks.

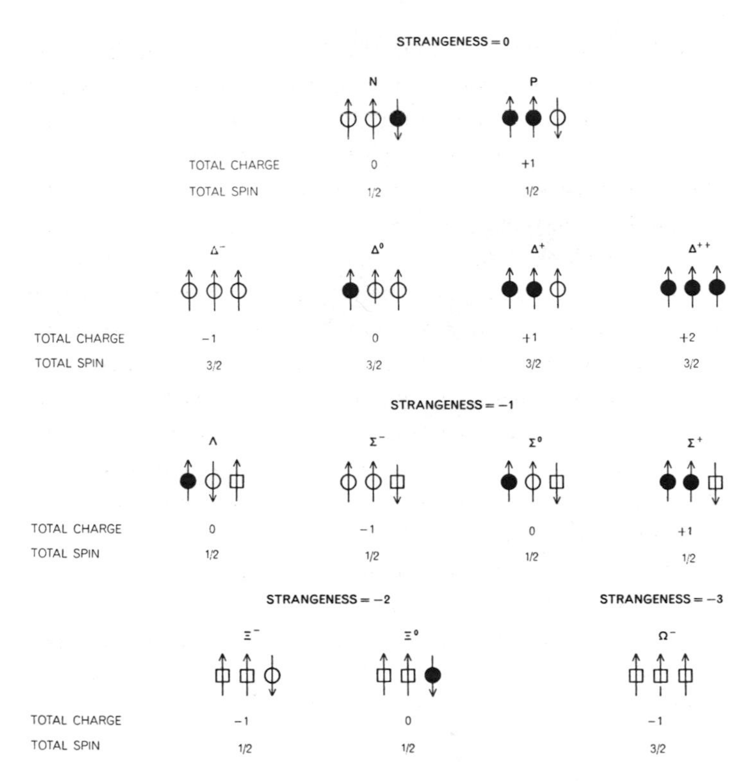

Figure 7.
Sets of three quarks in various combinations provide the known properties of baryons. When two spin arrows point in opposite directions, they cancel, thus the net spin of most of the baryons is 1/2. When all three arrows point the same way, the spin is 3/2.

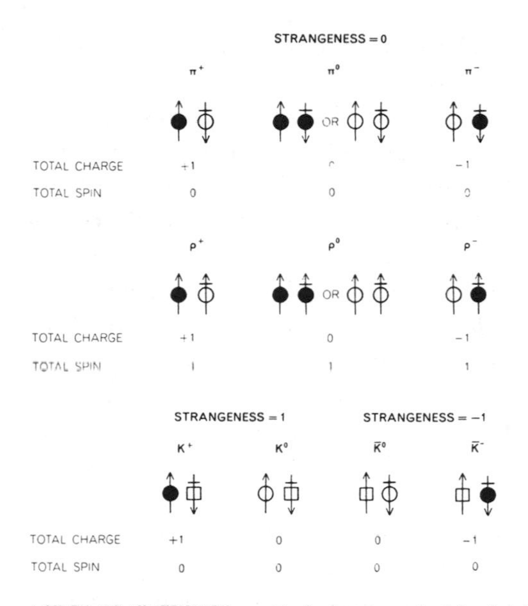

A QUARK AND AN ANTIQUARK account for the observed properties of the principal mesons. Pi (π) and rho (ρ) mesons have strangeness equal to zero, hence do not include a lambda quark. In the kaon sets, normal lambda quarks produce a strangeness of -1 for the K^0 and K^- pair and antilambda quarks produce a strangeness of $+1$ for K^0 and K^+.

Figure 8.
A quark and an antiquark account for the observed properties of the principal mesons. Pi (π) and rho (ρ) mesons have strangeness equal to zero, hence do not include a lambda quark. In the kaon sets, normal lambda quarks produce a strangeness of -1 for the K^0 and K^- pair and antilambda quarks produce a strangeness of $+1$ for K^0 and K^+.

Recently, theoretical physicists tried to combine not only isotopic spin and strangeness into a fundamental triplet but to include also the dichotomy of ordinary spin. The quarks are considered as particles with spin 1/2; instead of the three basic states of the quark, there are then six basic states, since each one may have the ordinary spin directed up or down. If the conditions are assumed to be invariant also to the direction of the ordinary spin, one obtains the so-called $SU(6)$ symmetry. The consequence of this wider symmetry is surprisingly well fulfilled. One finds, for example, that the quark-antiquark multiplets should exist with angular spin 0 and 1 and that the system of three quarks should have the total angular spin either 1/2 or 3/2. This is just what one finds in nature. In addition, most interesting predictions can be made in regard to magnetic moments; for the first time the ratio 3/2 between the proton and the neutron moment has been explained by a theory.

A more quantitative exploitation of these ideas resulted in even more surprising agreements with the observed facts. If one assumes, for example, that the dynamic conditions that hold these hypothetical quarks together are nearly invariant to an interchange of one of the three types of quark by another, then one obtains a super-multiplet structure for the levels of a three-quark system and of a quark-antiquark system. The formalism shows that the former system gives rise to singlets, octets, and decuplets, and the latter system to singlets and octets. For example, a baryon octet would consist of the proton, the neutron, the three Σ, the Λ, and the two Ξ. The decuplet would consist of the ten observed baryon states of angular momentum 3/2. One of the meson octets would consist of the three pions, the two kaons, the two antikaons, and the η-meson.

The assumption of complete equivalence of the three types of quarks would lead to the result that the members of these multiplets have all the same mass. In fact, this is not so; they differ in energy by far more than the members of isotopic spin multiplets which, in fact, are subgroups of the $SU(6)$ multiplets. However, the energy split within these multiplets follows a regular pattern, and it is just the pattern predicted by the $SU(6)$ formalism for the case in which a weak force existed that breaks the symmetry.

Roughly speaking, this split can be reproduced by the simple, symmetry-breaking assumption that the quark which carries strangeness is heavier than the other two. It would explain the general tendency of mass to increase with increasing absolute values of strangeness. Finer considerations make it possible also to reproduce the mass splits between multiplet members of equal strangeness. The accuracy with which these predictions are fulfilled is most striking and has led to the prediction and subsequent discovery of the baryon state of strangeness -3, the famous Ω particle. The predictions of approximate $SU(6)$ symmetry are not only confined to energy splits; it is also possible to predict certain quantitative relations between members of one multiplet with regard to other properties, such as magnetic moments, and transition probabilities. Also these predictions seem to be on the whole well fulfilled.

Altogether the quark model of baryons and mesons has been surprisingly successful in describing the properties of quantum states and the transitions between them. For example, let us consider the transitions that take place between the first excited state of the baryon, the delta (Δ) state, and the ground state. In the quark model the two middle states of the delta quartet (Δ^0 and (Δ^+) differ from the proton and neutron ground states only by virtue of the fact that in the deltas the three quarks have parallel spin, giving a total spin of 3/2 (1/2 for each quark), whereas in the proton one of the quarks has its spin opposed to the other two, so that the net spin is only 1/2. Hence a transition from Δ^0 and Δ^+ to the ground state corresponds to the spin flip of one quark. It is relatively easy to calculate the radiation emitted by such a flip if one knows the magnetic moment of the flipping quark, which can be figured out by assuming that the moments of the three quarks together should add up to the known magnetic moment of the proton. Such calculations have indeed given the observed electromagnetic radiation for the transition between delta and the proton with reasonable accuracy.

A remarkable aspect of this $SU(6)$ symmetry is the amalgamation in one symmetry of both the angular and the isotopic spin. This may lead to a deeper insight into the fundamental role of the spinor concept in our description of particles. So far, however, it has led to a number of difficulties. They come from the fact that the angular spin is inextricably connected with the orbital angular momentum. Remember that the angular momentum of a relativistic particle consists of spin and orbital parts that are combined in a different way in the so-called *large* and in the *small* components. Only in the nonrelativistic limit can one speak of a pure angular spin dichotomy. There is, of course, no such thing as an orbital momentum in the isotopic spin space. The analogy of isotopic spin and angular spin, which is based upon the dichotomy of basic states, breaks down under relativistic conditions.

It should be emphasized that quarks have never been observed, even in nucleon collisions of the highest available energy. It may be that the binding force that holds quarks together is so high that much larger energies are needed to tear them apart. It may also be that we are still far from understanding the structure of baryons and mesons and that we have found only a certain similarity between the properties of baryons and the behavior of a hypothetical three-quark system and between the properties of mesons and the behavior of quark-antiquark systems.

If the quark model were correct, we would have to assume the existence of a new and extremely strong force between quarks.

The nucleons would be "atoms" made up of quarks, and the force between nucleons—the nuclear force—would be analogous to the chemical force between atoms. Whatever the true nucleon structure may be, it is probable that the forces between nucleons are determined by the structure of the nucleons in some way. The complicated aspect of the nuclear force suggests to many physicists that it is an involved manifestation of something deeper that is actually simpler and more basic in character. One hopes that the nuclear force will one day be explained by some fundamental phenomenon in the internal dynamics of the nucleon, just as the complicated chemical forces today can be traced back to the simple electric attraction between nuclei and electrons.

Electrons and Weak Interactions

Thus Newton's question is not yet answered. We still ignore the basis for the unchanging properties of the nucleon. But more basic symmetries have been found, and we may be on the way to a deeper understanding of the reasons for the existence of the nucleon. However, we still owe Newton an answer as to the electron. The properties of the electron seem to be simpler in many ways, since it does not participate directly in the interplay of the newly discovered world of baryons and mesons. It interacts with the rest of the world only via the electric field and the weak interactions. But very little can be said to satisfy Newton's and our own curiosity with regard to the reasons why the electron has the properties that we observe. It is true that we understand better than ever the relations of the electron with the electromagnetic field. Quantum electrodynamics allows us to calculate all phenomena of this type with seemingly arbitrary accuracy. But this perfection is brought by abandoning any claims for understanding the charge and mass of the electron. They are "renormalized" to their experimental values. Being unable to explain them, we are still forced by ignorance to assume that these essential features are given to us *ab initio*. To make things worse, nature has provided us with a second kind of electron, the muon, which as far as we now know differs from the ordinary one only in its mass. The reasons for this duplicity are still totally obscure.

Even more mysterious are the roles played by the two electrons in the weak interaction. It is established today that all known particles interact by a universal weak interaction. The most characteristic feature of weak interaction process is that it is connected with an exchange of charge. When the electron interacts with a nucleon, it transfers its charge to the nucleon and assumes its uncharged state: It becomes a neutrino. The most common form of this process is a beta decay in which, say, a proton becomes a neutron and a positron-neutrino pair is emitted. Here the emission of a positron is equivalent to an absorption of an electron; hence the process corresponds to an encounter of an "incoming" electron with a proton, during which their charges are exchanged. The heavy electron behaves exactly like its lighter counterpart with respect to the weak interactions. It also possesses an uncharged form, the neutretto or muon neutrino, which now has been definitely proved to be different from the electron neutrino.

So far, not the slightest indication has been found of an internal structure of leptons. The electromagnetic interactions of both kinds of electron have not yet revealed any deviation from a point charge, and nothing like a spectrum of lepton states exists, except for the two forms, the charged and uncharged one.

The second striking feature of weak interaction is its violation of parity equivalence. Left-handed and right-handed processes are not equivalent. In fact, both kinds of neutrinos always appear with a spin opposed to their motion (left-handed screws). Until quite recently, this asymmetry was mitigated by the fact that antiparticles show exactly the opposite properties in their weak interactions; antineutrinos, for example, show up as right-handed screws; hence, weak interaction processes preserve an invariance if a left-right inversion is connected with particle antiparticle transformation (*CP* invariance). Lately, however, even this invariance has been put in doubt by experiments on *K* meson decays.

Baryons are subject to weak interactions in all quantum states. Strangeness or isotopic spin are not conserved in these interactions. It was most interesting to observe that the weak interactions with higher quantum states of the baryon are closely related to the weak interactions of the proton and the neutron. These relations bear out again the equivalence of the different baryon states on the basis of $SU(6)$ symmetry.

The weak interaction presents us with another fundamental problem: Our present understanding of interaction processes requires a field for the transmission of interaction, such as the electromagnetic field or the meson field. Does such a field exist for weak interactions? A recent search for the corresponding field quantum had negative results. Does this mean that the mass of the quantum is higher than the limit which could have been found with today's accelerators (about two proton masses), or that our ordinary field concepts do not apply to the weak interactions?

Summary

Let us try to summarize what we can answer today to Newton's question for the reasons of the unchanging properties of nature. The characteristic and well-defined structures of atoms and nuclei are based upon the symmetry of quantum states of these composite units. But the stability of their constituents is still poorly understood. There are two types of entities that we encounter here; they go under the names "lepton" and "hadron."

The leptons include the two electrons in their charged and uncharged form, and the hadrons include all baryons and mesons. To our knowledge, these entities are subject to mutual interactions of four different types, which we enumerate in the order of their strength: gravity, weak interactions, electromagnetism, and strong interactions. We omit gravity from our considerations, since its role in the world of elementary particles is completely unknown.

Weak interactions exist between all of these entities, leptons and hadrons alike; electromagnetic interactions are found between all particles carrying charges or magnetic moments; strong interactions exist only in the case of hadrons. Today we do not know whether the hadrons have an internal structure; hence it is not clear whether the strong interactions should be considered as acting between hadrons or as acting between the constituents of hadrons.

The symmetries of these interactions determine many of the properties of the entities and therefore are the essential shape-giving factors. It is most interesting that the number of symmetries increases with the strength of interaction. All of the interactions are subject to the translational and rotational symmetry of the space in which they are imbedded. This is a symmetry which appears to us as quite natural. The laws of conservation of energy, momentum, and angular momentum are a direct consequence of these symmetries. All interactions are subject also to two further conservations laws, which are not so well understood at present: the charge conservation, and the conservation of baryon and lepton number. Parity and strangeness, however, are not conserved by the weak interactions, but only by the electromagnetic interaction and those stronger than it; isotopic spin conservation holds only for the strong interactions; $SU(6)$ symmetry is valid for the strong interactions, but there exists a relatively weaker part of these interactions which violates it.

The stronger the interaction, the more symmetries exist. Is this remarkable fact significant for the ultimate explanation of the existence of elementary particles? It may be, for instance, that a certain number of symmetry principles imposes a unique dynamics, which then determines the properties of its fundamental units. It may also be that hadrons and leptons are not the ultimate structures at all; the hadrons may be composite structures of particles such as the quarks. If this were the case, the proton and the neutron would be a sort of "superatom" made up of fundamental particles; the nuclear force between the nucleons would be a kind of van der Waals force, an indirect effect of much stronger interactions within the superatom. Then the fundamental problem of elementary particles would reappear at a higher level when it is asked: Why do quarks exist? Most probably, however, the actual solution of the problem will take a new and wholly unexpected form.

It is fitting to close these remarks with another prophetic statement of Newton, the timeliness of which is almost uncanny:

> Now the smallest Particles of Matter may cohere by the strongest Attractions, and compose bigger Particles of weaker Virtue; and many of these may cohere and compose bigger Particles whose Virtue is still weaker, and so on for divers Successions, until the Progression ends in the biggest Particles on which the Operations in Chymistry, and the Colours of natural Bodies depend, and which by cohering compose Bodies of a sensible Magnitude.
> There are therefore Agents in Nature able to make the Particles of Bodies stick together by very strong Attractions. And it is the Business of experimental Philosophy to find them out. [2].

References

[1] I. Newton, *Opticks*, I. B. Cohen, ed., New York: Dover Publications, 1952, p. 400.
[2] Ibid., p. 394.

Part 3
Special Approaches

Part 3 contains articles dealing with widely differing subjects. "Nuclear Models" was written in 1950, shortly after the discovery of the nuclear shell model by Jensen, Mayer, and Haxel. This model explained many nuclear properties by assuming that the nucleons move almost freely and independently of each other within the nuclear volume, a most surprising assumption in view of the existence of strong nuclear forces between the nucleons. This article tries to explain in simple terms how free motion is possible in spite of strong interactions.

The second article, "The Visual Appearance of Rapidly Moving Objects," deals with the relativistic Lorentz contraction of rapidly moving objects. It describes the work of J. Terrell, who was the first to draw attention to the fact that the Lorentz contraction cannot be "seen" directly. On the contrary the optical picture of a rapidly moving object will appear not contracted but rotated in space. This remarkable fact had not been noticed for 55 years since Einstein's discovery of special relativity.

The third article, "How Light Interacts with Matter," deals with the interaction of light with matter, mainly in order to explain the appearance of objects when illuminated by visible light. It explains what happens when light falls upon atoms in gases, liquids, and solids; how the electrons move under the influence of the impinging light; and how they reflect or absorb the light which falls upon them. It describes how the different colors and shades of objects can be understood from their atomic properties.

The fourth article, "Fall of Parity," was written shortly after the discovery that certain fundamental processes in radioactivity show an asymmetry in respect to right- or left-handedness. It is an attempt to explain, in simple terms, the experiments that led to this discovery and to discuss some of the conclusions.

The last article, "The Role of Symmetry in Nuclear, Atomic, and Complex Structures," is a discussion of different forms of symmetry that occur in nature. It is based upon a lecture given at a Nobel symposium on symmetry in biological systems at the macromolecular level. It discusses the symmetry of larger molecules as being of different nature in contrast to the more fundamental symmetries of atomic states or subatomic particles. However, the former symmetry is based upon the latter; the symmetry of a flower is somehow predicated upon the fundamental symmetry of an atomic quantum state.

Nuclear Models

Since there exists no exact theory of nuclear structure, one is forced to introduce a number of oversimplified nuclear models in order to explain the main features of the experimental material. The models can be classified into two distinct groups according to their fundamental viewpoints: (*a*) the independent particle viewpoint (I.P.); (*b*) the strong interaction viewpoint (S.I.).

Recently the I.P. models have been widely discussed in connection with the surprisingly successful application of shell structure to nuclear properties [1]. One has observed abnormally large binding energies for nuclei for which either the neutron number or the proton number is equal to a series of so-called magic numbers. This phenomenon was interpreted by many authors by assuming that the nucleons move independently within a common potential trough. The energy levels in this trough are grouped in shells that are completely filled with particles (closed) when a "magic" number is reached. Very simple and general assumptions (e.g., spin orbit coupling) are sufficient to explain the observed values of the magic numbers. The physical properties of the different shells allow the prediction of more specific nuclear data, such as the occurrence of isomers, the spins, and, in some cases, the magnetic moments and the quadrupole moments of nuclei in their ground states.

It must be emphasized that this picture is based upon a far-reaching assumption: The nucleons must be able to perform several revolutions on their orbits before they are disturbed and scattered by the interaction with neighbors. This condition is necessary for the existence of a well-defined energy and angular momentum in each separate orbit. The "mean free path" within nuclear matter must be of the order of several nuclear radii in order to justify the existence of separately quantized independent states for each particle.

Revision by the author of an article originally in *Science* **113,** 2926, 1, 1951.

The S.I. models are based upon the opposite assumption. They are all derived from the concept of the compound nucleus. Bohr [2] has pointed out that, in most nuclear reactions, the incident particle, after entering the target nucleus, shares its energy quickly with all other constituents. This picture presupposes a mean free path of a nucleon that is much *shorter* than the nuclear radius. Nevertheless the compound nucleus picture is very successful in accounting for the most important features of nuclear reactions. To mention a few examples: the existence of closely spaced and narrow resonances in slow neutron reactions [2], the success of the evaporation picture of nuclear reactions with fast particles [3], the large values ($\sim \pi R^2$) of reaction cross sections with fast neutrons [4].

The two viewpoints seem to be totally contradictory. The nuclear forces as we know them from the deuteron and from two-particle scattering experiments represent a strong interaction and therefore suggest the validity of the S.I. viewpoint. In fact, the known scattering cross sections of elementary particles at 20–30 Mev (this is the order of the kinetic energy inside a nucleus) would indicate a mean free path of only 10^{-13} cm with nuclear matter. Hence the recent success of the I.P. shell model has led to speculations that envisage much weaker nuclear forces within a nucleus, compared to the ones observed between isolated pairs.

It is the purpose of this note to point out that the I.P. and the S.I. models are perhaps not as contradictory as it appears at first thought. It must be noted that the successful predictions of the shell model are always applied to the ground states or to the lowest excited states of nuclei. The regularities in the binding energies are properties of the ground state, the spin, and the magnetic and electric moments, too. The occurrence of isomers is a problem of the first excited state. Even the small neutron capture cross sections of "magic" nuclei [5] can be interpreted by assuming that the ground state of the target nucleus has an unusually low energy. Then the captured neutron forms a compound nucleus of an abnormally low excitation and its level density will also be abnormally low. This leads directly to a low capture cross section, since the neutron width is then much larger than the radiation width.

The applications of the S.I. models are restricted to problems involving high nuclear excitation. The compound nucleus formed in a nuclear reaction is always excited, at least to an energy larger than the binding energy of the added nuclear particle (about 8 Mev for protons or neutrons). Hence it seems that the strong interaction between nucleons within a nucleus is observed only at high nuclear excitations.

The failure of the S.I. viewpoint at low excitation energies does not necessarily imply that no strong interactions exist between nucleons. It is very probable that the Pauli principle prevents the strong interaction from exhibiting the expected effects. The interaction cannot produce the expected scattering within the nucleus, because all quantum states into which the nucleons could be scattered are occupied. Only at higher excitations, when not all of the lowest states are occupied, will scattering take place and prevent the formation of independent orbits.

It may be useful to discuss in this connection an analogous situation that one finds in the theory of the electron motion in solids. The electronic properties of metals and insulators can be described very successfully by assuming that the electrons move in a common potential field, the electric field of the ions in the lattice. The interaction between the electrons is completely neglected. The electronic states in the lattice field exhibit also a kind of shell structure, the Brillouin zones, and an insulator may be called a "magic" crystal for which the shells are completely filled.

The success of this description is perhaps also surprising in view of the fact that the interaction between electrons is by no means small. In fact, an electron with a few electron volts of energy that enters the metal from the outside is stopped in the metal within one or two interatomic distances, simply by the scattering with other metallic electrons. The mean free path of this electron within the metal is not greater than one interatomic distance, as can be shown with a simple calculation using the Rutherford scattering formula. In spite of this fact, the mean free path of the metallic electrons is very much greater than the interatomic distances; in fact, it is limited not at all by the interaction between the electrons but by the irregularities in the lattice. The reason is again found in the Pauli principle, which does not admit any scattering of electrons by electrons, because all states into which the scattering process may lead are occupied. This is not the case for the electron entering into the metal, since it possesses a surplus energy at least equal to the work function. Hence we find long mean free paths in the nonexcited state in spite of strong interaction, but short mean free paths in the highly excited state that is created when an electron enters from the outside.

The conditions of the electrons in a metal are obviously quite different from the conditions of the nucleons in a nucleus. There is no external field in the nucleus corresponding to the ionic field in the crystal. The common potential in the I.P. model is the average effect upon one single nucleon of all other constituents. However, the influence of the Pauli principle upon the mean free path of the electrons may serve as a useful analogy to understand the possibility of an I.P. picture in the presence of strong interaction between nucleons.

References

[1] W. J. Elsasser, *Phys. Radium* **5,** 625, 1934; M. G. Mayer, *Phys. Rev.* **74,** 235, 1948; *ibid.*, **75,** 1969, 1949; L. W. Nordheim, *Phys. Rev.* **75,** 1894, 1949; E. Feenberg, and K. Hammack, *Phys. Rev.* **75,** 1766, 1949; H. J. Suess, *Phys. Rev.* **75,** 1766, 1949.

[2] N. Bohr, *Nature* **137,** 344, 1936.

[3] H. Bradt and D. J. Tendam, *Phys. Rev.* **72,** 1117, 1947; E. L. Kelly and E. Segrè, *Phys. Rev.* **75,** 999, 1949; P. C. Gugelot, *Phys. Rev.* in press, 1950.
[4] H. Feshbach and V. F. Weisskopf, *Phys. Rev.* **76,** 1550, 1949.
[5] D. J. Hughes, W. D. B. Spatz, and N. Goldstein, *Phys. Rev.* **75,** 1781, 1949.

The Visual Appearance of Rapidly Moving Objects

I would like to draw the attention of physicists to a recent paper by James Terrell [1] in which he does away with an old prejudice held by practically all of us. We all believed that, according to special relativity, an object in motion appears to be contracted in the direction of motion by a factor $[1-(v/c)^2]^{1/2}$. A passenger in a fast space ship, looking out of the window, so it seemed to us, would see spherical objects contracted to ellipsoids. This is definitely not so according to Terrell's considerations, which for the special case of a sphere were also carried out by R. Penrose [2]. The reason is quite simple. When we see or photograph an object, we regard light quanta emitted by the object when they arrive simultaneously at the retina or at the photographic film. This implies that these light quanta have *not* been emitted simultaneously by all points of the object. The points further away from the observer have emitted their part of the picture earlier than the closer points. Hence, if the object is in motion, the eye or the photograph gets a "distorted" picture of the object, since the object has been at different locations when different parts of it have emitted the light seen in the picture.

In special relativity, this distortion has the remarkable effect of canceling the Lorentz contraction so that objects appear undistorted but only rotated. This is exactly true only for objects which subtend a small solid angle.

In order to understand the situation thoroughly let us consider the distortion of the picture we see of a moving object under nonrelativistic conditions, where light moves with light velocity c only in the stationary frame of reference of the observer, and a moving object does not suffer a Lorentz contraction. In the frame of the object moving with the velocity v the light velocity would be $c-v$ in the direction of motion and $c+v$ in the opposite direction.

Revision by the author of an article originally in *Phys. Today* **13,** 9, September 24, 1968.

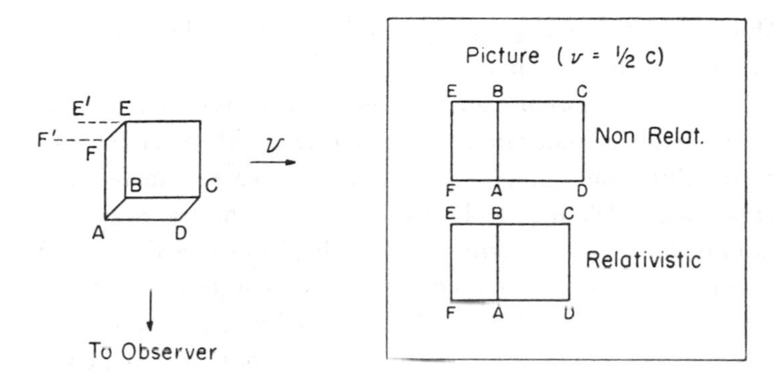

Figure 1.
A cube moving with velocity v seen by an observer at an angle of 90°.

We first consider the case of a cube of dimension l moving parallel to an edge and observed from a direction perpendicular to the motion. The observation is made at great distance in order to keep the subtended angle small (see Figure 1). The square *ABCD* facing the observer will be seen undistorted since all points have the same distance from the observer. The square *ABEF* facing in the opposite direction of the motion (the rear side in regard to the motion, not in regard to the observer's position) is invisible when the cube is not in motion. However when it moves it becomes visible since the light from E and F is emitted l/c seconds earlier when the points E and F were $(v/c)l$ further behind at E' and F'. Hence the face *ABEF* will be seen as a rectangle with a height l and a width $(v/c)l$. The picture of the cube, therefore, is a distorted one. In an undistorted picture of a rotated cube both faces should be foreshortened; if the face *ABEF* is shortened by the factor v/c, the other face *ABCD* should be foreshortened by $(1-v^2/c^2)^{1/2}$, whereas here *ABCD* appears as a square. Hence the picture of the cube appears dilated in the direction of motion. A similar consideration for a moving sphere shows that it would appear as an ellipse elongated in the direction of motion by a factor $(1+v^2/c^2)^{1/2}$.

We get even more paradoxical results by considering the picture of a moving cube in a nonrelativistic world, seen not at 90° to the direction of motion but at $180-\alpha$ degrees where α is a small angle. We now look at the object to the left when it is coming toward us from the left. We will assume now that $v/c=1$ in order to simplify our considerations. What is the picture then? Figure 2 illustrates the situation. The edges *AB*, *CD*, *EF* are denoted by the numbers 1, 2, 3. We assume that the edge 1 emits its light quanta at the time $t=0$. Where must the edges 2 and 3 emit their light such that it travels in a common front with the light from 1, in order to arrive simultaneously at the observer? It is easily seen that 2 must emit its light much earlier; in fact it must happen when it is at the point marked 2′ which is determined by the equality of the distances (2′2) and (2′*M*). The interval (2′2) is the distance which the edge 2 travels between the emission of light by 2 and 1. The length (2′*M*) is the

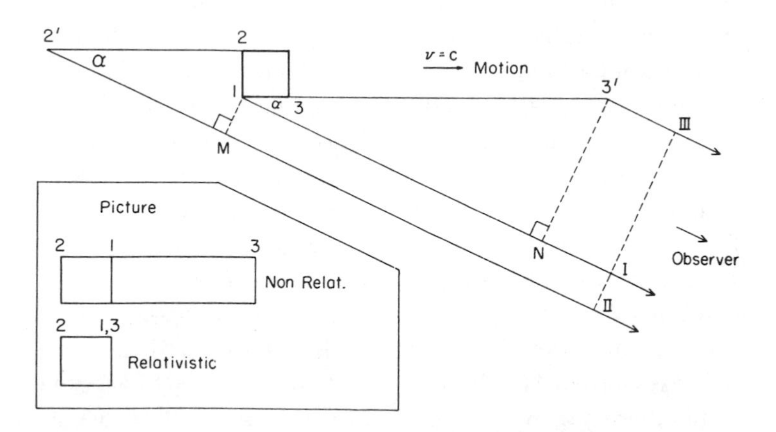

Figure 2.
A cube moving with light velocity viewed at an angle $\theta = 180° - \alpha$. Points I, II, III are light pulses coming from object points 1, 2, 3 and arriving simultaneously at the observer. (Nonrelativistic.)

distance which the light travels from 2′ in order to be "in li
with the light emitted by 1. Both light and edge travel with
speed c. We can see that the distance $(1M)$ is equal to $(1\,2)$ wh
is the size l of the cube. The light seen from edge 3 is emit
much later, when the edge is at 3′. The point 3′ is determinec
the equality of the distance $(3\,3')$ and $(1N)$. A simple calculat
shows that $(3'N) = l \sin\alpha\,(1-\cos\alpha)^{-1}$.

What then is the picture we see of the cube? It is indicatec
the figure by the points I, II, III which represent the positi
of the light quanta coming from the object and form the pictu
We will see a strongly deformed cube with the edge 1 in
middle, the edge 2 on the left of 1 as if we were looking fr
behind (from the left to the right) and the edge 3 quite far to
right of 1. Again we see a picture elongated in the directio
flight. The face between the edges 1 and 2 appears as a t
square.

We now will show that relativity theory simplifies the situati
It removes the distortion of the picture and what remains is
undistorted but rotated aspect of the object. We can see
directly with the examples quoted. Consider the cube w
looked at perpendicular to its motion: The Lorentz contrac
reduces the distance between the edges AB and CD by
factor $(1-v^2/c^2)^{1/2}$ and leaves the distance between AB and
unchanged. Therefore the picture of the face $ABCD$ is f
shortened precisely by the amount necessary to represen
undistorted view of a cube turned by an angle whose sine is
In the case of the cube moving with light velocity toward us
Lorentz contraction reduces the distance between the edg
and 3 to zero. The picture one sees then is a regular squ
corresponding to the rear face and nothing else, since ed
coincides with edge 1. Hence we see an undistorted pic
directly from behind. The object is undistorted but turnec
an angle of $(180-\alpha)$ degrees.

We can show by means of the following consideration that

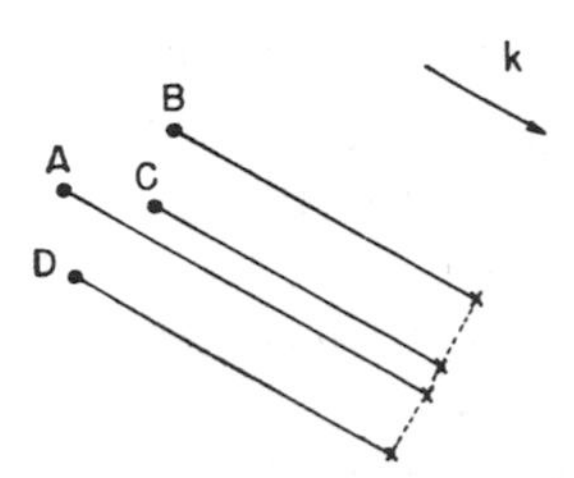

Figure 3.
A "picture" A, B, C, D are points of the object. The four crosses are the light pulses making up the "picture."

result is quite generally true for any object. Let us consider an assembly of light pulses originating from N points of the object, traveling all in the same direction given by a vector **k**, and such that the light pulses are all in one plane perpendicular to **k** (see Figure 3). Then they will arrive simultaneously at the eye of the observer and produce the shape which is seen. We will call such an assembly of light pulses a "picture" of the object. Under non-relativistic conditions a "picture" does not remain a picture when seen from a moving frame of reference. The reason is that, in the moving frame, the plane of the light pulses is no longer perpendicular to the direction of the propagation. In a relativistic world a "picture" remains a "picture" in any frame of reference. The light pulses would arrive simultaneously at a camera in every system of reference.

This fact can be proven immediately in the following way: The light pulses form a wave front or can be imagined as moving embedded in an electromagnetic wave exactly where this wave has a crest. It is known that electromagnetic waves are transverse in all frames of reference. That means that a wave front or the plane of the wave crest is perpendicular to the direction of propagation in *any* system. We can also show that the distance between the light pulses is an invariant magnitude. Here we only need to introduce a coordinate system where the x-axis is parallel

to the propagation. Then for two light pulses of the picture the invariant $(x_1-x_2)^2+(y_1-y_2)^2+(z_1-z_2)^2-c^2(t_1-t_2)^2$ is equal to the square of the distance d between the two pulses, since $d^2 = (y_1-y_2)^2+(z_1-z_2)^2$ and $x_1 = x_2$ when $t_1 = t_2$. The latter relation expresses the fact that the pulses are in a plane perpendicular to the propagation.

The only thing that is not invariant is the direction of propagation, the vector **k**. The transformation of this direction is given by the well-known aberration formula. A light beam whose direction includes the angle θ with the x-axis is seen including an angle θ' with the x-axis in a system moving with the velocity v along the x-axis:[1]

$$\sin\theta' = \frac{(1-v^2/c^2)^{\frac{1}{2}}\sin\theta}{1+(v/c)\cos\theta}.$$

We can conclude the following result from the invariance of the "picture": The picture seen from a moving object observed at the angle θ is the *same* as one would see in the system where the object is at rest but observed at the angle θ'. Hence we see an undistorted picture of a moving object but a picture in which the object is seemingly rotated by the angle $\theta'-\theta$. A spherical object still appears as a sphere.

This must not by any means be interpreted as indicating that there is no Lorentz contraction. Of course, there is Lorentz contraction, but it just compensates for the elongation of the picture caused by the finite propagation of light.

It is instructive to plot the angle θ' as a function θ. Figure 4 shows this relation for $v = 0$, for a small value of v/c, and also for the case $v/c \approx 1$. We see that the apparent rotation is always negative, which means that the object is turned such that it

[1]The angles refer to the direction in which the light beam is seen; that means a direction opposite to the motion of the light pulses.

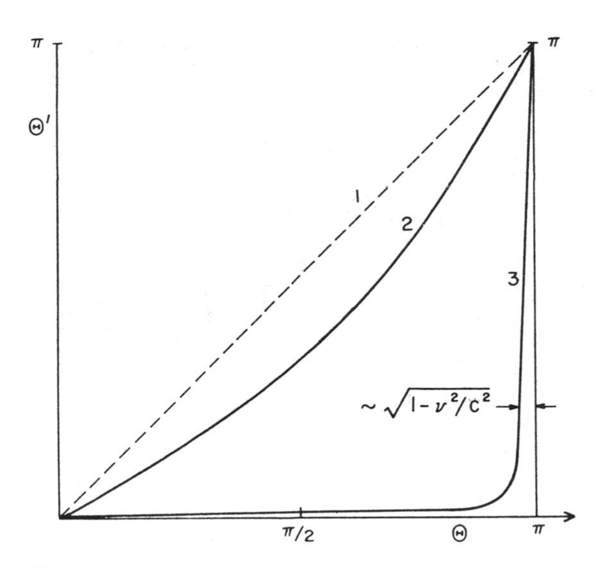

Figure 4.
The angle of observation θ' of a light beam relative to the direction of v, seen by an observer in the moving system, versus the same angle θ as seen in the rest system. Curve 1 is for $v = 0$, curve 2 is for $v = c/2$, curve 3 is for $v = c$.

reveals more of its trailing side to the observer. In the extreme case of $v \approx c$, θ' is extremely small for all values of θ except when $180 - \theta$ is of the order $[1-(v/c)^2]^{1/2}$. Since θ goes from 180° to 0° when an object moves by, we find for the case $v \approx c$ that we see the front side of the object only at the very beginning; it turns around facing its trailing side at us quite early when we still see it coming at us and remains doing so until it leaves us and naturally is seen from behind. This paradoxical situation is perhaps not so surprising when one is reminded of the fact that the aberration angle is almost 180° when $v \approx c$. Hence the light which we see coming from the object when it is moving towards us has left the object backwards when observed from the object itself.

The situation becomes clearer when we look closer at the distri-

bution of the emitted light as seen from the observer. Let us assume that the moving object emits radiation which is isotropic in its own rest system, i.e., its intensity is independent of the emission angle θ'. This radiation does not at all appear isotropic in the nonmoving system; there it seems concentrated in the forward direction. If $v \approx c$, most of the light appears to be emitted such that it includes a very small angle θ with v. This is a well-known effect which causes an isotropic emission to look as if almost all radiation is emitted in the form of a focused headlight beam. One example of this effect is the radiation of electrons running along a circle with a velocity near c as one finds it in synchrotrons. In this case the radiation in the rest system is not completely isotropic; it is essentially a dipole radiation. Still it appears as an emission sharply peaked in the direction of flight.

The apparent angular distribution $I(\theta)$ of the radiated intensity in the system at rest is connected with the angular distribution $I_0(\theta')$ in the system of the moving object by an expression which is related to the aberration formula:

$$\frac{I(\theta)}{I_0(\theta')} = \frac{\sin\theta' d\theta'}{\sin\theta d\theta} = K(\theta)$$

$$K(\theta) = \frac{1-(v/c)^2}{[1+(v/c)\cos\theta]^2}$$

where θ is the angle of observation, which means that the forward direction is near $\theta = \pi$. $K(\theta)$ is plotted in Figure 5 as a function of θ and we see that the width of the "headlight" beam is of the order $[1-(v/c)^2]^{1/2}$.

The factor $K(\theta)$ also determines the Doppler shift of the light. If the emitted light has the frequency ω_0 in the system moving with the object, the observer at rest sees a frequency $\omega = K^{1/2}\omega_0$. The frequency is increased or decreased by the square

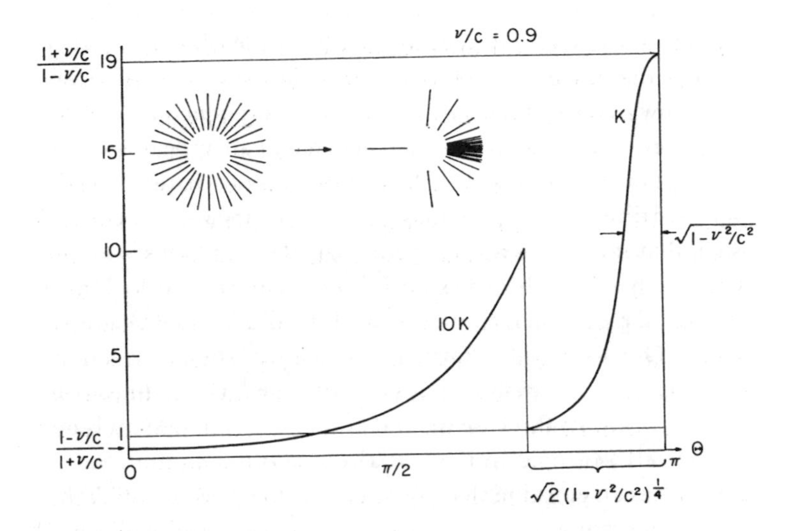

Figure 5.
The ratio K of the emitted intensity per solid angle measured in the observer's system to the intensity per solid angle measured in the system moving with the object, as function of the angle θ of observation. K also determines the Doppler shift. The observed frequency ω is related to the emitted frequency by $\omega = K^{1/2}\omega_0$.

root of the factor by which the intensity is enhanced or reduced. We note that for $\theta = 90°$ there is a reduction of frequency due to the relativistic Doppler shift.

We now describe what is seen when an object is moving by with a velocity near that of light. First, when the angle of vision is still near 180°, we see the front face of the object, strongly Doppler-shifted to very high frequencies and with high intensity. We are looking into the "headlight" beam of the radiation. When the angle of vision becomes of the order $\pi - [1-(v/c)^2]^{1/2}$ the color shifts towards lower frequencies, the intensity drops, and the object seems to turn. When $\theta \approx \pi - 2^{1/2}[1-(v/c)^2]^{1/4}$, still an angle close to 180°, the intensity becomes very low—we are out of the "headlight" beam—the color is now of much lower frequencies than it would be in the system moving with the object; the object has turned all around and we are looking at its trailing face. The front is invisible because the beams emitted forward in the moving system are concentrated into the small angle of the "headlight." The picture seen at angles smaller than $\pi - 2^{1/2}$ $[1-(v/c)^2]^{1/4}$ remains essentially the same until the object disappears. It is the picture expected when the object is receding. However it appears already when the object is moving toward us.

We would like to emphasize that all these considerations are exact only for objects which subtend a very small solid angle. Only then the picture consists essentially of parallel light pulses. If the angle subtended is finite we must expect different rotation angles for different parts of the picture and this would lead to some distortions. Moreover, it is correct only if the object is viewed with only one eye. A viewer using two eyes would observe a distortion because of the difference in angle between the two eyes. It has been shown by Penrose [2], however, that the picture of a sphere retains a circular circumference even for large angles of vision.

It is most remarkable that these simple and important facts of the relativistic appearance of objects were not noticed for 55 years until J. Terrell discovered and fully recognized them in his recent publication.

References

[1] J. Terrell, *Phys. Rev.* **116,** 1041, 1959.

[2] R. Penrose, *Proc. Cambridge Phil. Soc.* **55,** 137, 1959; H. Salecker and E. Wigner, *Phys. Rev.* **109,** 571, 1958.

How Light Interacts with Matter

The overwhelming majority of things we see when we look around our environment do not emit light of their own. They are visible only because they re-emit part of the light that falls on them from some primary source, such as the sun or an electric lamp. What is the nature of the light that reaches our eyes from objects that are inherently nonluminous?

In everyday language we say that such light is reflected or, in some cases, transmitted. As we shall see, however, the terms "reflection" and "transmission" give little hint of the subtle atomic and molecular mechanisms that come into play when materials are irradiated by a light source. These mechanisms determine whether an object looks white, colored, or black, opaque or transparent. Most objects also have a texture of some kind, but texture arises largely from the interplay of light and shadow and need not concern us here. I shall restrict my discussion mainly to the effect of white light on materials of all kinds: solids, liquids and gases.

White light, as it comes from the sun or from an artificial source, is a mixture of electromagnetic radiation with wavelengths roughly between 400 and 700 nanometers (billionths of a meter) and an intensity distribution characteristic of the radiation from a body that has a temperature of about 6,000 degrees Celsius. When such light impinges on the surface of some material, it is either re-emitted without change of frequency or it is absorbed and its energy is transformed into heat motion. In rare instances the incident light is re-emitted in the form of visible light of lower frequency; this phenomenon is termed fluorescence. In what follows I shall take up the commonest forms of secondary light emission. I shall undertake to answer such familiar questions as: Why is the sky blue? Why is paper white? Why is water transparent? What causes objects to appear colored? Why are metals shiny?

Revision by the author of an article originally in *Sci. Amer.* **219,** 3, 60, September 1968.

The answers are all based on the fact that the electrons of atoms are made to perform tiny vibrations when they are exposed to light. The amplitudes of these vibrations are extremely small: Even in bright sunlight they are not more than 10^{-17} meter, or less than 1 percent of the radius of an atomic nucleus. Nevertheless, all we see around us, all light and color we collect with our eyes when we look at objects in our environment, is produced by these small vibrations of electrons under the influence of sunlight or of artificial light.

What happens when matter is exposed to light? Let us go back to the simplest unit of matter and ask what happens when an isolated atom or molecule is exposed to light. Quantum theory tells us that light comes in packets called photons; the higher the frequency of the light (and the shorter the wavelength), the more energy per packet. Quantum theory also tells us that the energy of an atom (or a system of atoms such as a molecule) can assume only certain definite values that are characteristic for each species of atom. These values represent the energy spectrum of the atom. Ordinarily the atom finds itself in the ground state, the state of lowest energy. When the atom is exposed to light of a frequency such that the photon energy is equal to one of the energy differences between an excited state and the ground state, the atom absorbs a photon and changes into the corresponding excited state. It falls back to a lower state after a short time and emits the energy difference in the form of a photon.

According to this simple picture the atom reacts to light only when the frequency is such that the photon energy is equal to the difference between two energy levels of the atom. The light is then "in resonance" with the atom. Actually the atom also reacts to light of any frequency, but this nonresonant reaction is more subtle and cannot be described in terms of quantum jumps from one energy level to the other. It is nonetheless important, because most of the processes responsible for the visual appearance of objects are based on responses to nonresonant light.

Fortunately the interaction of light with atoms can be described rather simply. One obtains the essential features of that interaction—in particular the re-emission without change of frequency—by replacing the atom with electron oscillators. An electron oscillator is a system in which an electron is bound to a center with a spring, or some force equivalent to a spring, adjusted so that there is a resonance at the frequency ω_0. The electron oscillators we are using to represent an atom are designed so that their frequencies ω_0 correspond to transitions from the ground state to higher states. They represent the resonance frequencies of the atom in the ground state. Each of these oscillators has a certain "strength," a measure of the probability of the transition it represents. Usually the first transition from the ground state has the largest strength; that being so, we can replace the atom with a single oscillator.

Another quantity that characterizes these oscillators is their resistance coefficient, or friction. Friction causes a loss of energy in the oscillating motion. It describes a flow of energy away from the vibration into some other form of energy. It indicates that energy is being transferred from the excited state by some route other than the direct transition back to the ground state. Thus whenever the excited state can get rid of its energy by means other than re-emission of the absorbed quantum, the corresponding oscillator must be assumed to suffer some friction. This is an important point in our discussion, because excited atoms in solids or liquids transmit their excitation energy mostly into heat motion of the material. Unlike the isolated atoms found in rarefied gases, they have only a small chance of returning directly to the ground state by emission of a light quantum.

Henceforward I shall discuss the effect of light on atoms in terms of this oscillator model. We can now forget about photons and excited quantum states because one obtains correct results by considering the incoming light as a classical electromagnetic wave acting on classical electron oscillators. The effects of quantum theory are taken care of by the appropriate choice of oscillators to replace the atom. One can interpret the results of the oscillator model in such a way that under the influence of light the motion of the oscillators is superposed on the ordinary state of motion of the electron in the ground state. Whenever a light wave passes over the atom, a general vibration is set up in the ground state of the atom, a vibration of a kind and strength equal to the vibrations the oscillators of our model would perform if they were exposed to the light wave. The electron cloud of each atom vibrates under the influence of light. The cloud vibrates with the same frequency as the incoming light and with an amplitude corresponding to that of one of the model oscillators. It is this vibration, amounting to less than 10^{-17} meter in amplitude, that re-emits the light by which we see the objects around us.

The light from the sun or from artificial sources is a mixture of light of many frequencies. The motion of an oscillator exposed to such a mixture is simply a superposition of all the motions it would perform if exposed separately to light of each frequency contained in the mixture. Hence all one needs to know for the study of atoms under the influence of light is the motion of oscillators driven by an electric wave of a specific frequency.

If an electromagnetic wave of frequency ω passes over an electron oscillator, the electric field exerts a periodic force and leads to certain characteristic responses (see Figure 1). First of all the periodic electric field induces a vibration of the oscillator so that it oscillates with the frequency ω of the field, not with its own resonance frequency ω_0. The amplitude and the phase of this motion depend on the relative values of ω and ω_0. If ω is much smaller than ω_0, the oscillation is weak and in phase with the driving electric force of the light. If ω is much larger than ω_0, it is also weak but opposite in phase to the driving force. If ω is in resonance (in which case ω equals ω_0), the oscillation is strong and out of phase. This means: When the driving force is

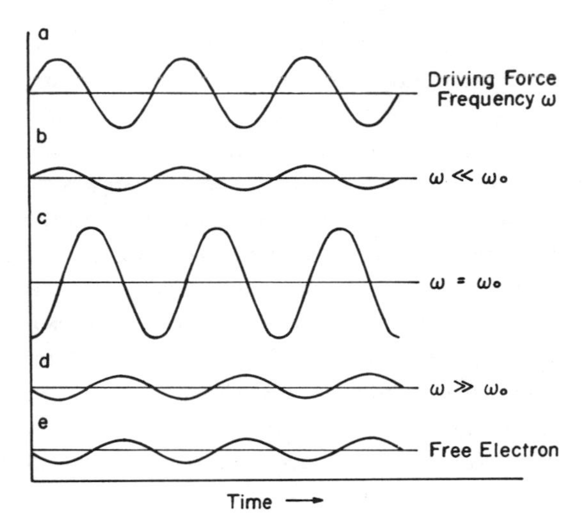

Figure 1.
Response of oscillator to periodic driving force serves as a model of how the electrons of an atom respond to the driving force of light. The response of each oscillator (*b*, *c*, *d*, *e*) depends on its particular resonance frequency ω_0. The driving force (*a*) has a frequency of ω. When ω_0 is much greater than ω, the oscillator responds in phase but only weakly (*b*). When ω_0 equals ω, the response reaches a maximum and is 90 degrees out of phase (*c*). When ω_0 is much less than ω, the response is again weak and 180 degrees out of phase (*d*). This weak response closely resembles the response of a free electron (*e*).

at its crest, the oscillation goes through the zero point. The amplitude of the oscillation follows a fairly simple mathematical formula that need not concern us here. The formula shows that if ω is much smaller than ω_0, the amplitude is small but is almost independent of the driving frequency ω. If ω is much larger than ω_0, the amplitude decreases with increasing ω at a rate proportional to $1/\omega^2$. Only the resonance case $(\omega = \omega_0)$ corresponds to the simple picture of a transition to another quantum state.

What are the resonance frequencies in different atoms and molecules? Most of the simple atoms such as hydrogen, carbon, oxygen, and nitrogen have resonances with frequencies higher than visible light; they lie in the ultraviolet. Molecules, however, can perform vibrations in which the atoms move with respect to one another within the molecule. Because of the large mass of the nuclei, such vibrations have very low frequencies; the frequencies are lower than those of visible light, in the infrared region. Hence most simple molecules such as O_2, N_2, H_2O, and CO_2 have resonances in the infrared and ultraviolet and no resonances in the visible region. They are transparent to visible light. Nevertheless, visible light has an influence on them, which can be described by our oscillator picture. We replace the molecules by two kinds of oscillator, one representing the ultraviolet resonances, the other the infrared resonances. The latter are not really electron oscillators; they are "heavy" oscillators in which the mass of the oscillating charge is as large as the mass of the vibrating atoms, since they are supposed to represent the motions of atoms within the molecule.

We are now ready to understand one of the most beautiful colors in nature: the blue of the sky. The action of sunlight on the molecules of oxygen and nitrogen in air is the same as the action on the two kinds of oscillator. Both oscillators will vibrate under the influence of visible sunlight. The amplitude of the infrared oscillators, however, will be much smaller than the amplitude of the ultraviolet oscillators because of their higher vibrating mass. Accordingly we need to consider only the oscillators with ultraviolet resonance. When the oscillators are under the influence of visible sunlight, the force that drives them is below the resonance frequency. Therefore they vibrate with an amplitude that is roughly equal for all visible frequencies (see Figure 2).

We must now take into account the fact that a vibrating charge is an emitter of light. According to a principle of electrodynamics an electron oscillating with an amplitude A emits light in

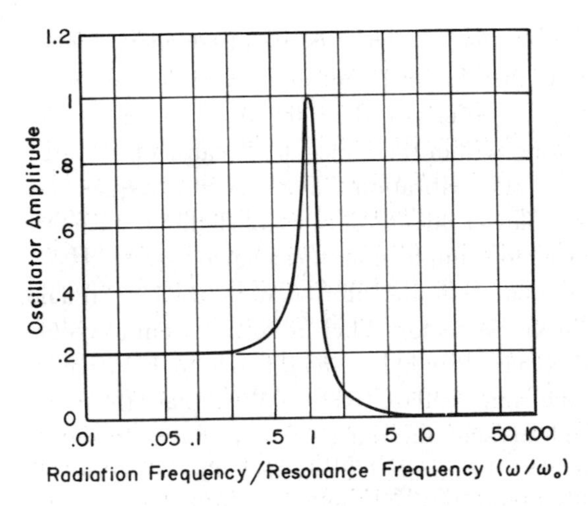

Figure 2.
Oscillator amplitude is a function of the ratio between the frequency of the driving force ω and the oscillator's resonance frequency ω_0. This ratio ω/ω_0 is expressed in the horizontal scale, which is logarithmic. The amplitude approaches a constant value (left) when the driving frequency is much below resonance. This is the situation when molecules of nitrogen and oxygen in the atmosphere are exposed to visible light. When the driving frequency is much above resonance, the amplitude decreases as the square of ω/ω_0.

all directions with an intensity given by a formula in which the intensity of the radiation is proportional to the fourth power of the frequency. (The formula is $1/3(e^2/c^3)\omega^4 A^2$, where e is the charge of the electron, c the velocity of light, and ω the frequency of oscillation.) Hence the molecules of air emit radiation when they are exposed to sunlight. This phenomenon is known as Rayleigh scattering. It is called "scattering" because part of the incident light appears to be diverted into another direction. Whenever we look at the sky but not directly at the sun, we see the light radiated by the air molecules that are exposed to sunlight. The scattered light is predominantly blue because the reradiation varies with the fourth power of the frequency; therefore higher frequencies are re-emitted much more strongly than the lower ones.

The complementary phenomenon is the color of the setting sun. Here we see solar rays that have traveled through the air a great distance. The higher-frequency light is attenuated more than light of lower frequency; therefore the reds and yellows come through more strongly than the blues and violets. The yellowish tint of snowy mountains seen at a distance is a similar phenomenon. The stronger attenuation of higher frequencies is a consequence of the conservation of energy: The energy for the reradiation must come from the incident sunlight, and because there is stronger reradiation at the higher frequencies more energy is taken from the sunlight at higher frequencies.

In actuality Rayleigh scattering is a very weak phenomenon. Each molecule scatters extremely little light. A beam of green light, for example, goes about 150 kilometers through the atmosphere before it is reduced to half its intensity. That is why we can see mountains at distances of hundreds of miles. Lord Rayleigh exploited the phenomenon of light scattering to determine the number of molecules in a unit of volume in air. In 1899 he was admiring the sight of Mount Everest from the terrace of his hotel in Darjeeling, about 100 miles away, and he concluded from the dimness of the mountain's outline that a good part of its light was scattered away. He determined the scattering power of each molecule from the index of refraction of air, and he found the number of air molecules per cubic centimeter at sea level to be 3×10^{19}, which is very close to the correct value.

Now we know why the sky is blue. Why, then, are clouds so white? Clouds are small droplets of water suspended in air. Why do they react differently to sunlight? The water molecule also has resonances in the infrared and in the ultraviolet, resonances not much different from those of oxygen and nitrogen molecules. Water molecules should react to sunlight in a similar way. There is, however, an essential difference. We determined the scattering of sunlight in air by assuming that each molecule reradiates independently of the others, so that the total scattered intensity is the sum of the individual molecular intensities. That is correct for a gas such as air because gas molecules are located at random in space, and thus there is no particular interference among the individual radiations of the molecules in any direction other than the direction of the incident sunlight.

That is no longer the case when the molecules or atoms take on a more orderly arrangement, as they do in solids and liquids and even in the droplets of a cloud. In order to understand the effect of light on matter in bulk, we must study how electromagnetic waves react to a large number of more or less regularly arranged oscillators, when the average distance between the oscillators is small compared with the wavelength of visible light. As we have seen, under the influence of incident light every oscillator emits a light wave. Because the oscillators are no longer randomly spaced, however, these waves tend to interfere with one another in a definite way: There is constructive interference in the forward direction (the direction of the light path) and destructive interference in all other directions. The individual waves build up to a strong wave called the refracted wave; in any other direction the waves tend to cancel one another. If the oscillators are in a regular array, the cancellation is complete (Figure 3).

The refracted wave travels with a velocity v that differs from the ordinary light velocity c. The ratio c/v is called the refractive index n of the medium. There is a simple relation connecting the value of n with the amplitude A of the oscillator vibrations. The greater this amplitude is under the influence of a given and fixed driving force, the more n departs from unity. Knowing the refractive index of air, Lord Rayleigh used this relation to find the amplitude in sunlight of the oscillators representing the air molecules.

In a regular and uniform arrangement of atoms the re-emitted waves build up to a single refracted wave. There is no individual, or incoherent, scattering by each oscillator, as occurs in a gas

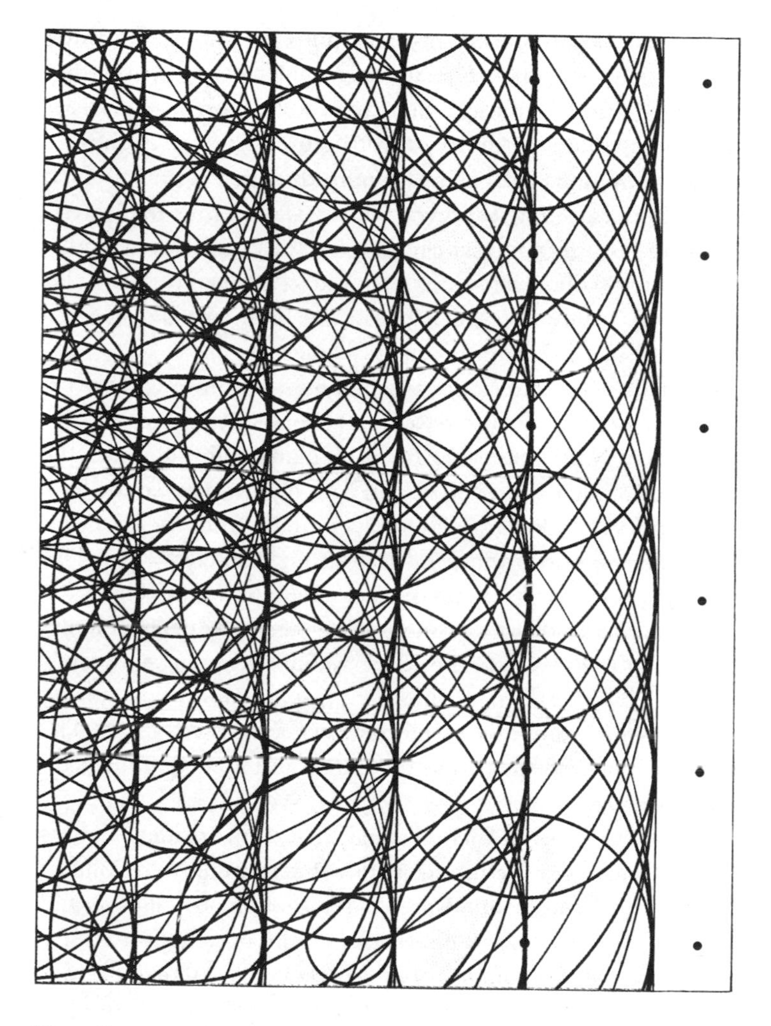

Figure 3.
Refracted wave in crystal consists of a parallel series of plane waves formed by the crests of many spherical waves (circles). The spherical waves depicted here represent light reemitted by the atoms (dots) in a crystal that is exposed to a light beam from the left. The refracted wave is traveling to the right. The situation in glass and water is similar.

such as air. As long as a crystal or a liquid contains heat it cannot be completely regular. The atoms or molecules are constantly vibrating, and in addition there are always some irregularities and imperfections in the crystal structure. These irregularities scatter some of the light away from the direction of the refracted wave. This scattering, however, is much weaker than the scattering in air, assuming equivalent numbers of atoms. For example, water is 1,000 times denser than air, but its incoherent molecular scattering is only five times greater per unit volume than the scattering in air that makes the sky blue.

Now let us see what happens when light impinges on the surface of a liquid or a solid or a cloud droplet. Again we replace each atom by an oscillator. These oscillators vibrate under the influence of the incident light and emit light waves. In the bulk of the material all these light waves, apart from the weak incoherent scattering, add up to one strong refracted wave. This is not so, however, near the surface of the material. There is a thin layer of oscillators at the surface (about as deep as half a wavelength) for which the black radiation is not completely canceled by interference since there are no atoms beyond the surface whose radiation would produce that interference. The radiations backward of these oscillators add up to a "reflected" wave (Figure 4).

What is the color of this reflected light if the incident light is white? One might perhaps conclude that it should be as blue as the sky, since it too comes from the reradiation of oscillators and we have learned that the intensity of this reradiation is proportional to the fourth power of the frequency. Actually it is as white as the incident light. The intensity of the reflected light with respect to the incident light in water, glass, or crystals is practically independent of the frequency.

The explanation is that the reflected wave is a coherent com-

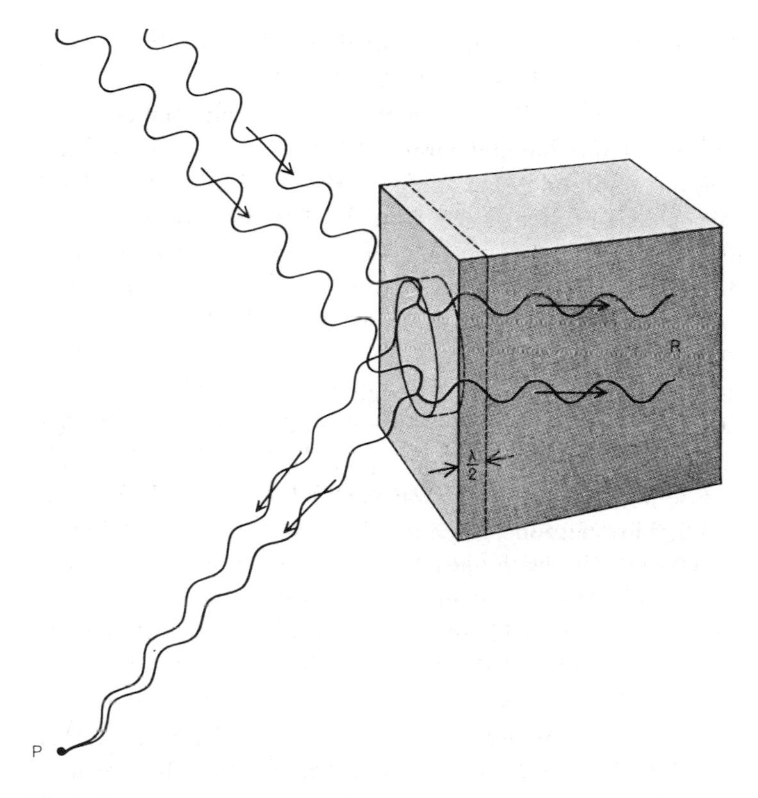

Figure 4.
Reflection of light from the surface of a solid or liquid involves only the oscillators (electrons) located in a small, pillbox-shaped volume at the surface of the material. When light (I) impinges on a smooth surface, part of the light proceeds into the material as a refracted wave (R) and part is reflected toward the observer (P). The radiation that makes up the reflected wave originates in a thin layer whose thickness is about half the wavelength of the incident light. The oscillators whose radiation adds up coherently at P are contained in a flat cylinder whose top surface is about λd in area, where λ is the wavelength of the light and d is the distance from the surface to the observer. This area is called the first Fresnel zone. For a spherical surface of radius R the area of the first Fresnel zone is equal to $\lambda \pi R$, provided that the distance to the observer is large compared with R.

posite of many individual reradiations. The oscillators, since they are not randomly distributed, reradiate in unison. That by itself would not yet explain the difference; it would only tell us that the reradiated intensity is high. In a coherent radiation it is the amplitudes that add up and not the intensities. Hence N coherent oscillators give N^2 times the intensity of one individual radiation. It still would seem that the reradiation should be blue, inasmuch as the radiation intensity of each oscillator increases strongly with frequency. What happens, however, is that the number of oscillators acting in unison also depends on the frequency: The layer that gives rise to reflection is half a wavelength deep, and the area of the layer whose reflected light arrives in step, or with the same phase, at a given point in space is also proportional to the wavelength. (This area is known as the first Fresnel zone. The radiations from all other parts of the surface interfere with one another, so that they give no light at that point.) Hence the number N of oscillators producing light in unison is proportional to the square of the wavelength. The intensity of this light is proportional to N^2. The net effect is to cancel the fourth power of the frequency, because higher frequency means smaller wavelength and a smaller value of N. As a consequence the reflected-light intensity is independent of frequency. Therefore clouds are white: The incident sunlight is reflected at the surface of the water droplets without change in spectral composition.

On the same basis we can understand the transparency of water, of glass, and of crystals such as salt, sugar, and quartz. If light impinges on these substances, it is partially reflected at the surface but without preference for any color. The rest of the light enters the substance and propagates as a refracted wave within it. Therefore these objects look colorless. Their outlines are nonetheless visible because of the reflection of the light at the surfaces.

Sometimes such objects may exhibit color under special circumstances. Reflection and refraction are only approximately independent of frequency. Both increase slightly at higher frequencies because such frequencies are a little closer to the natural resonance of the atom. Although these differences amount to only a few percent, they can become important if the details of refraction and reflection are critically involved in the way the light returns to the observer. Then, as in the case of a rainbow, these small differences may spread white light into its constituent colors.

Transparent substances with a large smooth surface reflect part of the incident light in a fixed direction according to the familiar laws of reflection. Therefore extended plane surfaces of colorless substances (windowpanes, water surfaces) can produce mirror images. If such colorless substances are in the form of small grains, each grain being larger than the wavelength of light, the substances appear white, like clouds. The incident white light is partially reflected in many directions, depending on the orientation of the grain surfaces. The light that pentrates the grains is again partially reflected at the inside surfaces, and after several reflections and refractions it comes back to the eye of the observer from various directions. Since none of these processes discriminates against any color, the returning light will be white and diffuse. This explains the color of snow, of salt and sugar in small grains, and of white pills and powders: All consist of small crystals of molecules with resonances only in the infrared and in the ultraviolet. The whiteness of paper has the same origin. Paper consists of an irregular weave of transparent fibers (Figures 5 and 6). The molecules of the fibers also have no resonances in the visible region. The fibers reflect and refract light in the same way as fine grains of salt or snow.

If the grains are smaller than the wavelength of light, there are not enough oscillators in the grain to establish ordinary reflection and refraction. The situation is then more as it is in a gas of independent molecules, and the substance looks bluish. One can see this on a dry day when a cloud disappears. What often happens is simply that the droplets become smaller and smaller by evaporation until the cloud appears blue. The blue color of cigarette smoke is also evidence that the particles are smaller than the wavelength of visible light. The color of the sky above our cities is largely determined by the way sunlight is scattered by particles of smoke or dust, some larger than the wavelength of light, some smaller. That is why the city sky is a pale mixture of white and blue—far from the deep, rich blue that prevails where the air is clear.

Although water is transparent because it has strong resonances only in the infrared and the ultraviolet, it does have a slight color of its own. This is not the wonderful deep blue one often sees on the surface of a lake or an ocean. That blue is the re-

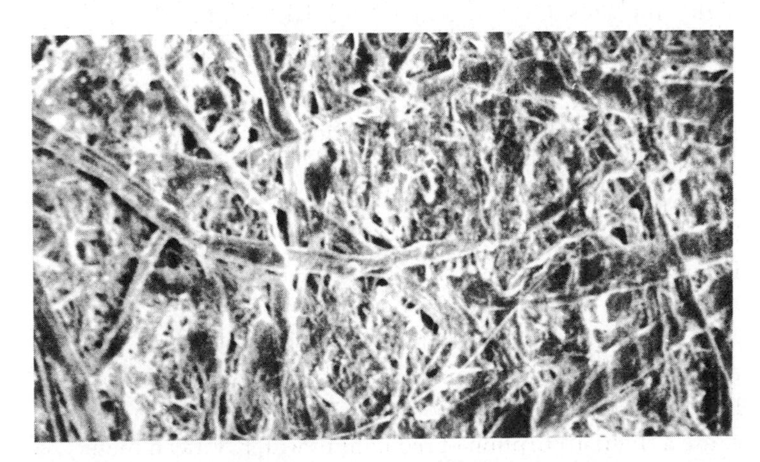

Figure 5.
Ordinary paper consists of a random mesh of translucent cellulose fibers. This 125-diameter magnification was made with a scanning electron microscope by Consolidated Papers, Inc. In such a micrograph the object appears to be tilted at an angle of 45 degrees.

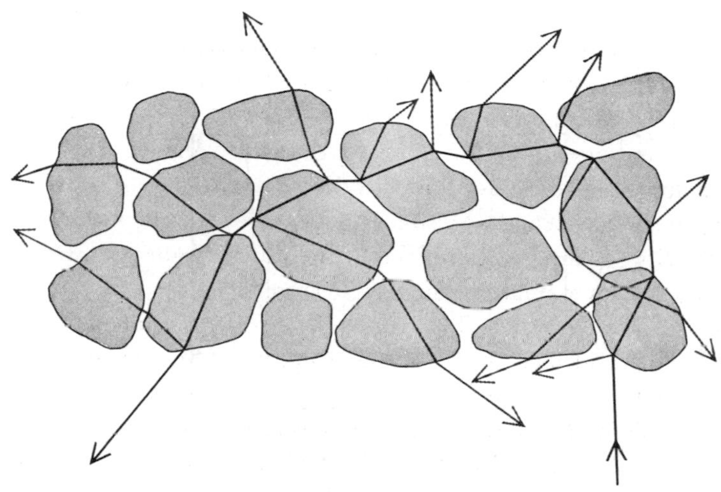

Figure 6.
Reflection of light from paper surface involves many refractions and reflections as the rays of incident light perform a random walk through a mesh of translucent fibers, represented here in cross section. The multiple refractions of a single entering beam are traced, beams reflected from various surfaces are shown.

flected color of the sky. The intrinsic color of water is a pale greenish blue that results from a weak absorption of red light. Because of its strong electric polarity the water molecule vibrates readily when it is exposed to infrared radiation. Indeed, its infrared resonances are so strong that they reach even into the visible red (Figure 7).

These resonances represent true absorptions of light because the energy of the light quantum absorbed is transformed into heat motion. The weak resonances in the visible red therefore cause a slight absorption of red light in water. Fifteen meters of water reduces red light to a quarter of its original intensity. Practically no red sunlight reaches a depth below 30 meters in the sea. At such depths everything looks green. Many deep-sea

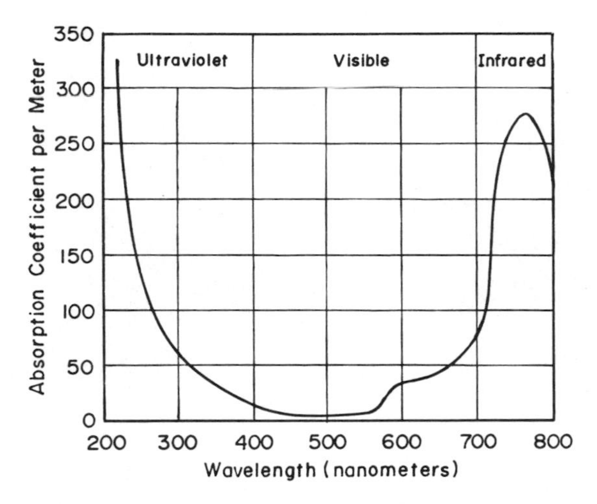

Figure 7.
Light absorption by water is negligible between 400 and 580 nanometers in the visible part of the spectrum. The absorption increases in the orange and red region and rises steeply in the near infrared. Absorption is also strong in the ultraviolet. The absorption is caused by resonances of the water molecule in response to various wavelengths.

crustaceans are found to be red when they are raised to the surface. In their normal environment they appear black. The selection mechanisms of evolution could not distinguish between black and red under such conditions.

The greenish-blue color of water is different in kind from the blue color of the sky. It is a color produced by the preferential absorption of the red and not by the preferential re-emission of the blue, as it is in the sky. One way to be convinced of this difference between air and water is to look at a white object under the surface of water: It looks bluish green. On the other hand, a snowy slope seen through many miles of air looks yellowish. In the first instance the red light was absorbed; in the second the blue light was scattered away.

Most of the colors we see around us are due to preferential

absorption: the colors of leaves, flowers, birds, butterflies, rubies, emeralds, and the whole gamut of paints and dyes. What accounts for the preferential absorption in such a diverse range of things and substances? Most atoms and molecules have resonances only in the infrared and the ultraviolet. In order to produce a resonance in the visible region the excitation energy must be between 1.5 and 3 electron volts. These are rather small values for electron excitations and large values for molecular vibrations. There are, however, atoms and molecules that do have excited states in that region. They are atoms with several electrons in incomplete shells and certain organic compounds: the dyestuffs. Such atoms can be excited by rearranging the electrons in the incomplete shell, which requires less energy than excitation to a higher shell. The dyestuffs are chain or ring molecules in which the electrons move freely along the chain or the ring. They are spread out, so to speak, over larger distances than electrons in ordinary atoms and molecules. The excited states in such a system are of lower energy than they are in atoms, because larger size gives rise to longer electron wavelengths, and this in turn is associated with lower frequency and thus lower excitation energy. Thousands of chemists have devoted their professional lives to the synthesis of organic molecules that have resonances in one part or another of the visible spectrum (Figure 8).

Although low-lying excited states give rise to resonance frequencies in the visible region, other conditions must be fulfilled before a molecule will serve as a dye. First, one must be sure that the light quantum is not simply re-emitted after its absorption has lifted the molecule into the excited state. One wants the energy of the excited state to be transformed preferentially into heat motion. This will be the case if we deal with matter in bulk, liquid or solid. Under such circumstances re-emission of light is very improbable. Second, the resonance frequencies must be

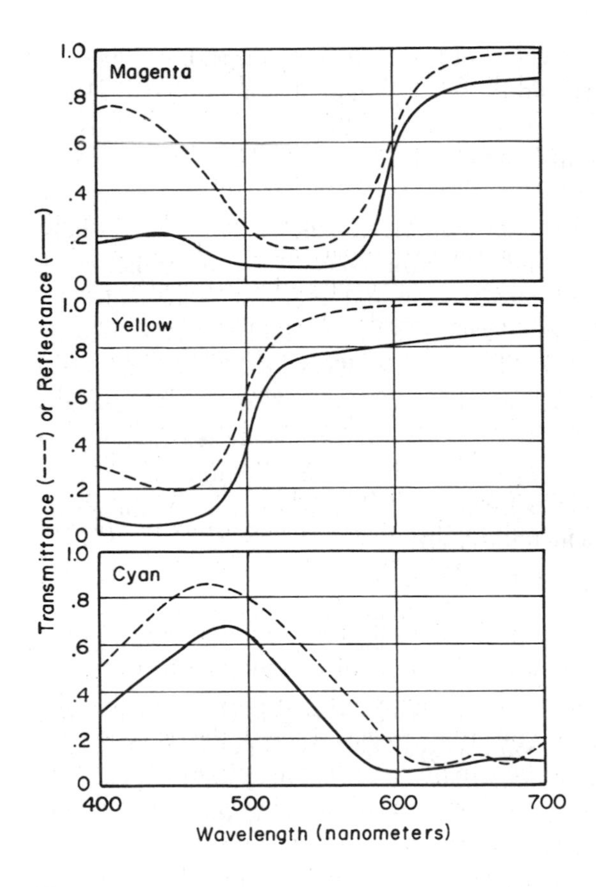

Figure 8.
Spectral characteristics of colored inks (solid curves) and of an optimum set of dyes for color film (broken curves) are plotted in these three panels. The dyes of color film can produce colors that are more highly saturated than those attainable with printing inks. The reason becomes clear in these curves: each of the color film dyes transmits more of the desired wavelengths than the corresponding printing ink is able to reflect.

spread over a broad interval. A dye with a narrow absorption band would reflect most wavelengths and thus look practically white. Here again matter in bulk contributes to the desired effect. In liquids and solids the energy levels of atoms or molecules are expanded into broad energy bands, with the result that resonances are spread over broad ranges of frequency. For example, a red dye absorbs light of all visible frequencies except the red. A green paint absorbs red and yellow as well as blue and violet. The absorption of a dye covers the visible spectrum with the exception of the actual color of the material. Some people may have wondered why a mixture of paint of all colors gives rise to a dirty black, although we are told that white is the sum of all colors. Colored paints function not by adding parts of the spectrum but by subtracting them. Hence a mixture of red, green, and blue paints will absorb all wavelengths and look virtually black.

A simple and striking color effect is the one produced by a stained-glass window. The dyestuff is contained in the glass. When light falls on stained glass, it is partially reflected at the surface, just as it is by ordinary glass. Indeed, the reflection is a little stronger for those frequencies that are absorbed, because, as we saw earlier, the amplitude of vibration is larger when the frequency is in resonance with the system. This effect, however, is usually not very pronounced, since the main reflection comes from the oscillators with resonances in the ultraviolet, as it does in ordinary glass. The part of the light that penetrates the body of the glass—the refracted wave—is subjected to the absorbing effect of the dye. Accordingly only light of the frequency that is not absorbed will pass through the glass. That is why one obtains such impressive color effects when white light penetrates stained glass. The color of the glass is less strong when one looks at the side that is illuminated. The reflection from the surface is practically colorless; the principle color one sees is from light that has penetrated the glass and is reflected again by the second surface (Figure 9).

A painted sheet of paper will serve as an example of ordinary painted objects. The paint causes the fibers of the paper to become impregnated with dye. When white light falls on paper, it is reflected and refracted many times before it comes back to our eyes. Whenever the light penetrates a fiber, the dye absorbs part of it: The fibers act as small pieces of stained glass. The best color effect is achieved when the reflecting power of the fiber is not too strong, so that most of the light enters the fiber. One remembers childhood experience with watercolors, which are most intense while the paper is still wet. The water reduces the

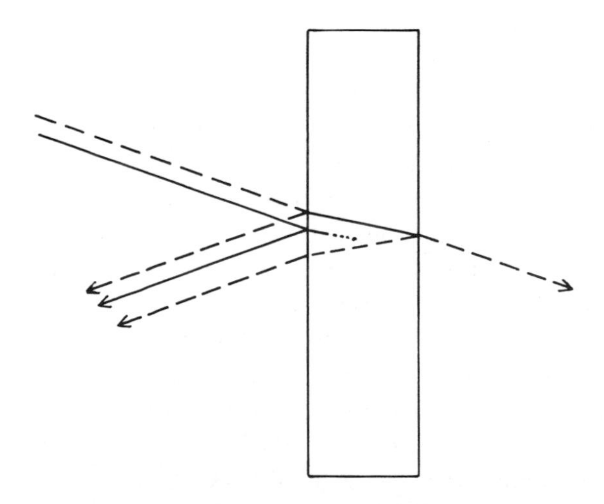

Figure 9.
Color of stained glass depends on which wavelengths the glass absorbs. Here it is assumed that the glass absorbs the shorter wavelengths so that primarily red light is transmitted. Thus blue light enters the glass and is absorbed, it is also partially reflected. Red light is reflected from the rear surface as well as from the front, so that the total reflected light is predominantly red. Thus red stained glass looks red by reflected as well as by transmitted light; the transmitted color is purer, however.

difference in refraction between the fibers and the interstices, thereby reducing the reflection of the fiber surfaces.

Glossy colored paper has a smooth surface. Its irregularities are small compared with the wavelength of light. The incident light is partially reflected without much preference for one color over another, but it is reflected by the smooth surface at a fixed angle according to the familiar laws of reflection. At any other angle most of the light that reaches the eye will have made several passages through the fibers before leaving the paper. That is why the color of the light is clear and deep: It is free of any uncolored direct reflection. Sometimes glossy paper demonstrates the fact that being in resonance implies larger amplitudes and therefore stronger reflection. One often notices an increased reflection of the deeply colored parts of a glossy picture, but only when the dye is deposited in the uppermost layer. Examine the illustrations in this article.

Objects are black when there is absorption for all visible frequencies. Well-known examples are graphite and tar. Black objects do not absorb all light that falls on them. There is always some reflection at the surface. Think of the reflection of the polished surface of a black shoe. A dull black surface reflects as strongly as a polished one but the reflected rays are distributed in all directions. A black surface with very low reflectivity can be produced by placing a few hundred razor blades in a stack. When the edges of the blades are viewed end on, they appear to be nearly dead black even though they are highly polished. The explanation is that light is trapped between the closely spaced edges and is absorbed after being reflected many times.

The most beautiful colors of all—the colors of plants, trees, and flowers—are based on the same principle of preferential absorption. The cells of plants are filled with dyes: chlorophyll in green leaves and blades of grass, other dyestuffs in the petals of flowers. White light that falls on plants is reflected and refracted by the cells; a large part of the light enters the cells, in the same way it does the fibers of paper. When it returns to the eye, all the colors but one or two are strongly reduced by absorption. Only green light escapes from chlorophyll-containing cells, only red light from the petals of red flowers.

We now turn to the visual appearance of metals. A metal is characterized by the fact that within the confines of the material there are many electrons—the conduction electrons—moving almost freely over many atomic diameters. These electrons are most important for the optical properties of metals. There is one, two, or sometimes three electrons per atom among the conduction electrons. The rest of the atomic electrons remain bound to the atoms. The conduction electrons can be regarded as an electron gas that penetrates the crystal lattice without much hindrance. The reason for this quasi-free motion lies in the wave nature of the electron. Although it is true that an electron wave is scattered by each of the metal atoms, the regular arrangement of atoms in the lattice makes the scattered waves interfere in a definite way: The waves all add up to one undisturbed wave in the forward direction. The corresponding electron motion is therefore the motion of a free particle. This is a phenomenon closely related to the formation of a refracted wave when light penetrates a crystal.

The motion of the conduction electrons is not completely free, however. Thermal agitation of the crystal lattice and other lattice imperfections produce some scattering away from the main electron wave. This is closely analogous to the weak scattering of the refracted light wave in a crystal. The effect can be expressed as a kind of friction of the conduction electrons. It is the cause of the electrical resistance in metals. In the reaction of electrons to visible light, however, friction does not play an important role; we are allowed to consider the electrons as freely moving.

What is the behavior of a free electron under the influence of light? It performs vibrations of the same frequency as the frequency of the driving force but of opposite phase. When the force is moving in one direction, the electron moves in the other one. It is the same kind of movement an oscillator performs when the driving frequency is much higher than its resonance frequency. A free electron in plain sunlight performs vibrations with an amplitude about 10 times larger than the amplitude of electrons in water and in crystals, or several times 10^{-17} meter.

What happens when light impinges on a metallic surface? The answer is, very much the same as happens when light strikes the surface of a liquid or a crystal, but there is one important difference. Since the resonance frequencies of a liquid or a crystal are higher than the frequency of light, they vibrate in phase with the light. In a metal, however, the electrons vibrate in *opposite* phase. Under these conditions a refracted light wave cannot be propagated if the density of electrons and the amplitude of their vibration is above a certain limit. The limit can be expressed in terms of the "plasma frequency" ω_p, which is given by the equation $\omega_p = (\mathcal{N}e^2/m)^{\frac{1}{2}}$, where $\mathcal{N}$ is the number of electrons per cubic centimeter and m is the electron mass. This frequency usually is in the ultraviolet. Whenever the light frequency is less than ω_p, as it always is for visible light, no refracted wave can develop in the medium; there are too many electrons inside moving in phase opposite to the light.

Since no light energy can propagate into the material, all energy of the incoming light must go into the reflected wave. Just as with water or glass, the reflected light wave is produced in a thin layer at the surface of the metal, a layer no thicker than the wavelength of the light. A more exact calculation shows that in a metal this thickness is equal to the wavelength corresponding to the plasma frequency divided by 2π. This value is less than 10^{-7} meter. Unlike the wave reflected from water and glass, however, the wave reflected from a metal surface has almost the full intensity of the incoming wave, apart from small energy losses due to the friction of the vibrating electrons in the surface. This is why "white" metals such as silver and aluminum are so shiny: They reflect almost all visible light regardless of its frequency. Smooth surfaces of these metals therefore are ideal mirrors.

In colored metals such as copper or gold there are additional losses apart from the electron friction. These losses come from absorption by electrons other than the conduction electrons. Each atom in a metal is surrounded by shells of those electrons that remain with the atoms after the conduction electrons have detached themselves and formed the gas of free electrons. The resonance frequencies of the remaining electrons are usually in the ultraviolet and thus do not contribute to any color. In copper and gold, however, the bound electrons are part of an incomplete shell and do have resonances in the blue-violet that lead to absorption. As a result copper and gold have a reddish-yellow appearance.

Many color phenomena have not been treated here: the color of thin films, of fluorescent materials, of light emitted by flames, of electric discharges as produced in neon tubes, and many others. We have taken up only the most common features of colored objects in order to provide some insight into the optical processes that occur on the surface of things when they are illuminated by light and seen by our eyes.

Fall of Parity

A number of recent experiments in nuclear physics have revealed that some of the very basic properties of nature seem to be different from what we believed them to be. It is rare in the history of physics that the results of only a few experiments force upon us a change in our fundamental principles. This is just what has happened now, and this essay tries to explain the situation.

Before describing the experiments themselves, we will discuss the basic principle which is attacked by their results. It is the *principle of parity*. This principle can be stated in the following form: Any process which occurs in nature can also occur as it is seen reflected in a mirror. Thus nature is mirror-symmetric. The mirror image of any object is also a possible object in nature; the motion of any object as seen in a mirror is also a motion which would be permitted by the laws of nature. Any experiment made in a laboratory can also be made in the way it appears as seen in a mirror, and any resulting effect will be then the mirror image of the actual effect. In more elegant language, the laws of nature are invariant under reflection.

As an example, take a perfectly uniform bar supported in the middle by a pivot, as in Figure 1. We all know that it will not tip, but let us prove this using mirror symmetry, or the principle of parity. There are three possibilities: (i) The bar could tip clockwise, (ii) it could tip counterclockwise, or (iii) it could remain horizontal. Suppose we place a mirror as in Figure 2 (the dotted line represents the mirror). The mirror image of the bar and its support is identical with the object. However, if motion (*a*) were the correct one, the mirror image would show motion (*b*) and not the correct motion (i); hence, we have a contradiction to the principle of parity. Only the possibility (iii) is identical with it

With L. S. Rodberg. Revision by the author of an article originally in *Scie*
125, 3249, 627, April 5, 1957.

reflection and thus must be the correct one since the object itself is identical with its reflection.

Electromagnetic Radiation

Let us now look at a more sophisticated example. We will examine the radiation from an electric dipole. Such a dipole can be pictured, for example, as a charge which oscillates up and down in the z-direction (Figure 3). We see, first of all, that the radiation pattern will be uniform around the z-axis. This is because the electric dipole exhibits a cylindrical symmetry about this axis.

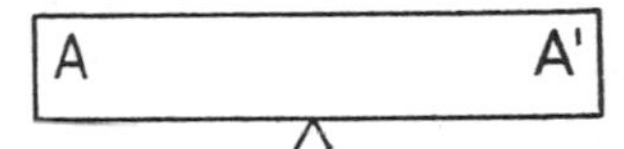

Figure 1.
Uniform bar pivoted in the middle.

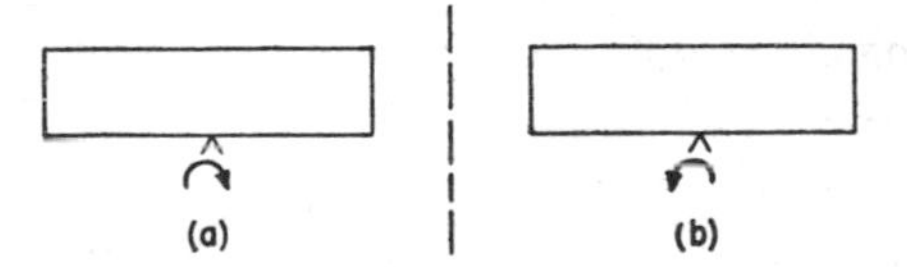

Figure 2.
Mirror image of bar. This illustrates the lack of mirror symmetry if the bar should be tilted.

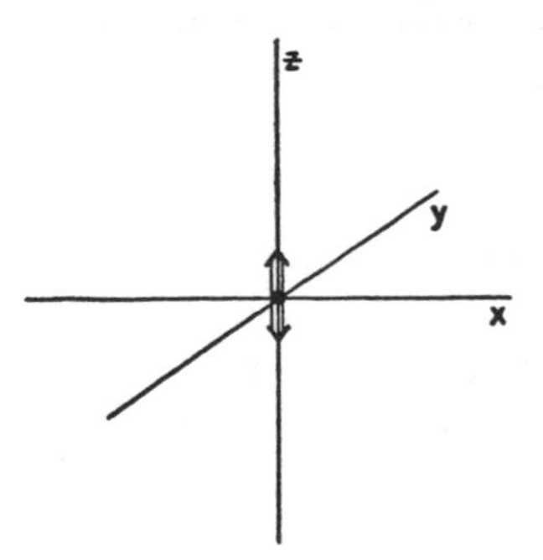

Figure 3.
Oscillating dipole.

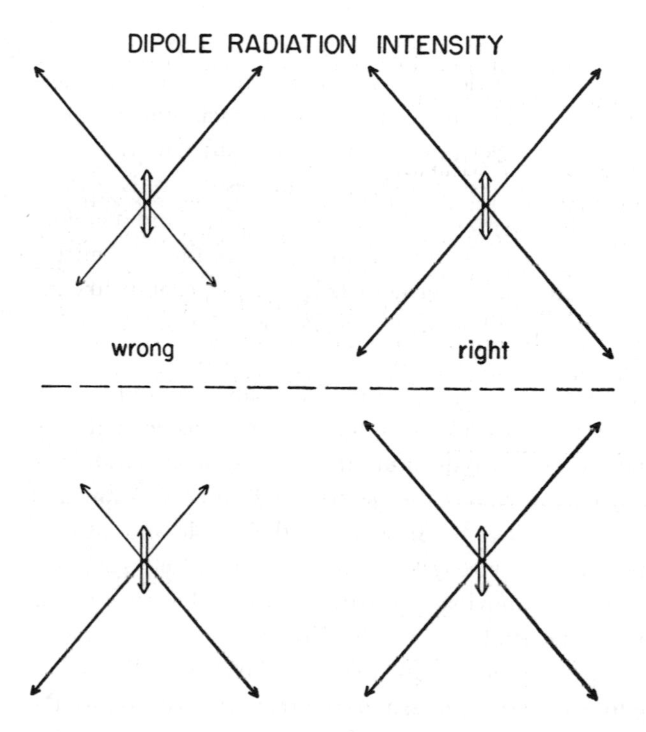

Figure 4.
Dipole radiation-intensity patterns and their mirror images. The pattern marked "right" emits the same intensities into the upper and lower hemispheres; the pattern marked "wrong" emits more into the upper hemisphere than into the lower.

We will now use mirror symmetry to show that the intensity is the same above and below the x-y-plane. In Figure 4 we illustrate the two cases. The mirror image of the oscillating dipole is identical with the object, apart from a phase shift of half a period. When the object moves up, the image moves down. However, the radiation intensity pattern is constant in time, and therefore it is not affected by this shift in time. We see that the mirror image of the radiation pattern labeled "right" is exactly like the actual one, as it should be, while the pattern labeled "wrong" is inverted: The object has a stronger field downward, while the image has a stronger field upward. They cannot both be right.

Let us now look at the electromagnetic field associated with this radiation. Here we examine the instantaneous position of the moving charge, and of the electric field, since we know that after each half period the direction of the dipole and also the direction of the field strength change their sign. Let us suppose that the charge is moving upward. We know that the electric field must be perpendicular to the direction of propagation at large distances, because light waves are transversal. We would like only to decide the question of the relative directions of the electric field in two beams, one going upward and the other downward. In fact, we want to decide between the two possibilities marked "wrong" and "right" in Figure 5. Using mirror symmetry, we can rule out the possibility marked "wrong." This situation cannot hold, for the dipole is turned around in the reflected picture, while the electric fields have the same directions as before reflection. (Alternatively, if we wait half a period, the mirror dipole will point upward again, but the electric field will have reversed its direction.) On the other hand, in the situation marked "right," the electric field has "followed" the dipole upon reflection. (Here, if we wait half a period, the mirror dipole and electric field will reverse, reproducing the present actual dipole and electric field.) Then the situation marked "right" must be the true one.

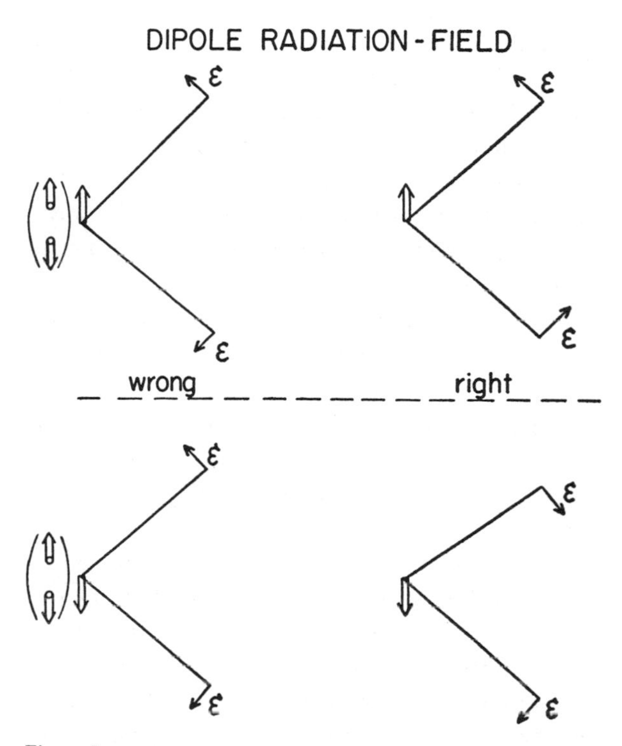

Figure 5.
The electric field distribution of a dipole radiation and its mirror image. The little arrows marked ϵ represent the direction of the electric field at a given time in the wave emitted in the direction shown. The heavy arrow in the center shows the displacement of the dipole at that time. The two heavy arrows in parentheses symbolize a quadrupole source.

On the other hand, a quadrupole consists of two dipoles opposite each other. It is thus unchanged when it is reflected in a mirror, so that we have the reversed case; the electric field of quadrupole radiation must be invariant upon reflection in a mirror, and the case marked "wrong" in Figure 5 would be the correct one. We say that dipole radiation has an "odd" parity since E has changed direction; quadrupole radiation has an "even" parity, since E is unchanged.

We have used the electric field in this discussion since it alone specifies a direction (the direction of the force on a positive charge). This direction becomes the reflected direction when seen in a mirror. The magnetic field does not specify a direction, but only a sense of rotation (for example, the sense of rotation of a moving charged particle which produced it). However, the sense of rotation is unchanged under reflection. It is important to remember here that the "direction" of the magnetic field is usually defined in terms of an arbitrarily chosen "right-handed screw." That is, we associate the magnetic field with a screw, which arbitrarily ascribes a direction to a sense of rotation in order to express it by a vector. This situation is usually described by saying that the electric field is described by a polar vector which changes direction under reflection, while the magnetic field is described by an axial vector, which does not change direction under reflection.

Let us now consider an object such as a screw which has a "helicity"—that is, a direction of motion associated with a sense of rotation (Figure 6). Its mirror image has the opposite helicity and must also exist in nature, by our principle of parity. Thus, in Figure 6 we see that we may place our testing mirror in two positions, one of which reflects the direction of motion but leaves the sense of rotation unchanged, while the other has the reverse effect. In either case, the helicity is changed.

An example is the tetrahedral molecule of Figure 7. We see

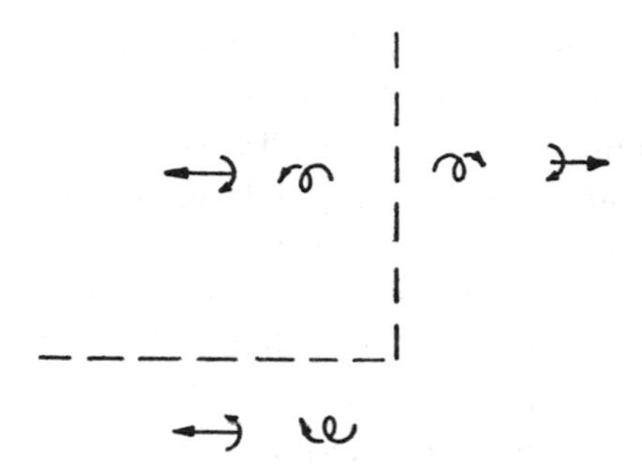

Figure 6.
A spiral and its mirror images. The horizontal mirror changes the direction but not the sense of rotation. The vertical one changes the sense of rotation but not the direction.

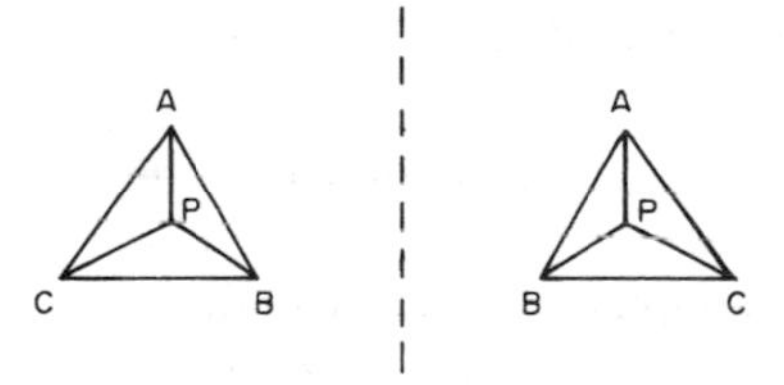

Figure 7.
An asymmetric tetrahedral molecule viewed in a mirror. The center point P is located above the plane of ABC.

that the reflected molecule cannot, by any rotation, be made to be identical with the original molecule (just as we cannot turn our left hand in such a position that it looks like our right hand). Thus these are distinct molecules which, by the principle of parity, must both exist in nature. An example of this situation is the quartz crystal, composed of many of these molecules. This crystal illustrates on a large scale this "handedness." The principle of parity requires that both types of crystals be found in nature.

A well-known example is the fact that sugar occurs in two varieties. However, it is only the right-handed kind, glucose, which is found in living matter. As physicists, we do not believe that this indicates an inherent handedness of nature; rather, we believe that it can be attributed to an accident which occurred at the origin of life. Life could just as well have developed by using levose instead of glucose.

Beta Decay

We now proceed to consider the actual experiments which have shed new light on this principle of parity, in particular, experiments on beta decay. All we need to know here is that there are atomic nuclei which emit electrons along with neutral, massless particles known as neutrinos. For instance, the isotope of cobalt known as cobalt-60 becomes nickel-60 and emits an electron (e^-) and an antineutrino ($\bar{\nu}$)

$$Co^{60} \rightarrow Ni^{60} + e^- + \bar{\nu}$$

The cobalt nucleus has a spin—that is, it is rotating with a well-defined angular momentum when it is in its normal state. Now we ask: In what directions will the electrons emerge? In a normal piece of cobalt, electrons will emerge in all directions because nuclei are oriented in all directions because of the heat motion.

Suppose we orient the nuclei—that is, force all the nuclei to align their axes of rotation parallel to a given direction and have them rotate in the same sense. This is the difficult part of the experiment since it is so hard to "get hold of" the nucleus. The only way is through the magnetic moment arising from the spin. The spin can be forced into a given direction by an external magnetic field if we can reach temperatures of less than 0.1 °K. Then it is possible to orient the nuclei.

What do we now expect? The nuclei are all rotating in the same sense. Let us apply the principle of parity. In a mirror

(Figure 8) they rotate the same, but the direction of the electrons is reversed. Thus the situation marked "wrong," in which more electrons emerge in one direction than in the other, violates the principle of parity: The mirror image contradicts the actual situation. Since the parity principle requires both to be right, we must exclude this case. Hence, we expect the same number of electrons to emerge in each direction.

This now sets the scene for the experiment. It was performed at the National Bureau of Standards in Washington, D.C., where the cryogenic equipment was available for experimenting at very low temperatures. The physicists who did it were C. S. Wu from Columbia University and E. Ambler, R. W. Hayward, D. D. Hoppes, and R. P. Hudson of the National Bureau of

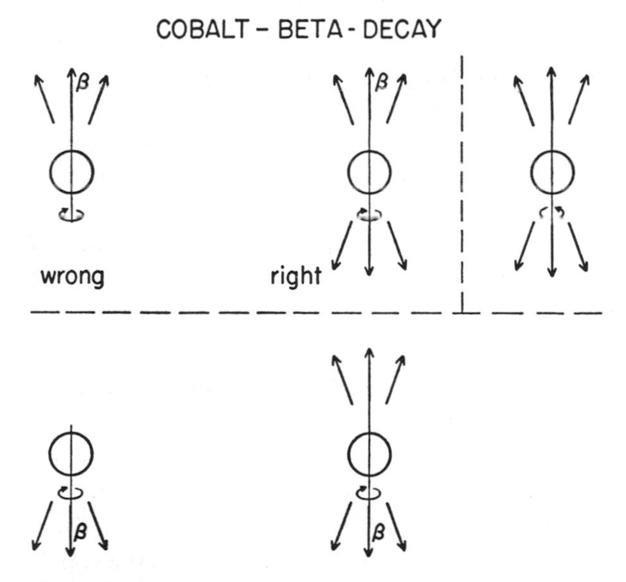

Figure 8.
Cobalt beta decay. Possible electron decay patterns and their mirror images. Only the choice which is mirror-symmetric should occur if the principle of parity is valid.

Standards. They oriented the rotation of cobalt nuclei and compared the electron intensities in the two opposite directions along the axis of rotation.

There are several remarkable features about this experiment. It is one of those experiments which only a few people would perform because the result "obviously" follows from mirror symmetry. Great discoveries are always made when one doubts the "obvious." In this case, it was the insistence of two theoretical physicists, T. D. Lee of Columbia and C. N. Yang of the Institute for Advanced Study, which prompted the experimenters to look for the effect. Lee and Yang suspected that the principle of parity may be invalid for certain weak interactions like beta decay.

Another remarkable feature of this experiment is the size of the effect which was measured. The intensity of electrons in one direction along the axis of rotation was found to be 40 percent larger than it was in the other. It is very rare in the history of physics that the failure of an established principle shows up with such large effects in the first experiment. Usually the first doubts are based on small deviations which hardly exceed the limits of error, and only after the passing of time and the application of great effort by many people would the effect be substantiated.

In view of the historic importance of this experiment, it is perhaps worthwhile to show the actual curves as measured. They are reproduced in Figure 9. The scale labeled "time" is actually a scale of temperature. The cobalt sample is cooled to a temperature at which its nuclei are aligned, and then it slowly warms up in the course of time. The curve labeled "gamma anisotropy" really tells us the fraction of nuclei which are oriented. For a large anisotropy, most of the nuclei are aligned. As the cobalt warms up, the heat motion causes the alignment to become more random, and the gamma anisotropy decreases.

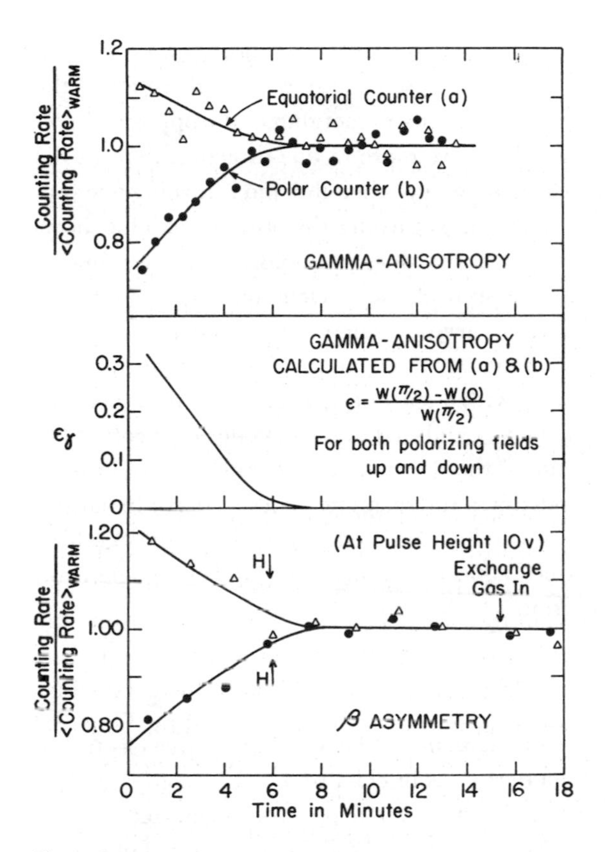

Figure 9.
Experimental observations on β-decay of cobalt-60 (Wu, Ambler, Hayward, Hoppes, and Hudson). The gamma anisotropy measures the orientation of the nuclei. The β-asymmetry measures the number of electrons which emerge parallel, and antiparallel, to the magnetic field.

The curve labeled "β asymmetry" is the significant one. This tells us the number of electrons emerging in the direction of the magnetic field, and the number emerging in the opposite direction. We see that there are more in one direction than in the other, that the electrons go up when the spin is turning one way and down when the spin is turning the other way. This shows that the principle of parity does not hold in this experiment. Remember that the spin of the nucleus tells only a sense of rotation. And yet the electron emerges in a preferred *direction*. This is the mark of the parity violation. The fact that there is a direction associated with a sense of rotation shows that there is a definite "handedness" exhibited in the beta decay of cobalt-60. The mirror image of the decaying cobalt nucleus would have the opposite handedness and seemingly does *not* occur in nature.

The same experiment has also been done with cobalt-58, which is a *positron* emitter. It goes over into iron-58 and emits a positron (e^+) and a neutrino

$$Co^{58} \rightarrow Fe^{58} + e^+ + \nu$$

where ν denotes the neutrino. Whenever a negative electron is emitted in a beta decay, as in cobalt-60, it is accompanied by an antineutrino, and whenever a positron is emitted, it goes with a neutrino. Most significantly, the same group of physicists have found the opposite handedness in the positron case. For the same rotational sense of the nucleus, negative electrons seem to emerge in one direction and positrons in the other.

Helicity

A possible explanation of these new phenomena has been proposed by Lee and Yang, and independently by L. Landau in Moscow and by A. Salam in England. They suggest that the helicity is associated with the neutrino, since all other phenomena in nuclear physics, which involve no neutrinos, exhibit perfect mirror symmetry. With this hypothesis, the difficulty is isolated from the rest of physics. It "minimizes the damage" and puts this strange property on the neutrino, which is already a strange particle.

Lee, Yang, Landau, and Salam argue that the neutrino is a spiral. Its sense of rotation and its direction of propagation are connected such that they form, say, a left-handed screw. The neutrino has the property that its spin (its rotation) must be such that its axis is parallel to its motion and its sense such as to form a left-handed screw. The antineutrino is supposed to have the opposite properties. It forms a right-handed screw.

It is interesting to note that particles with such properties must always move with the velocity of light c and, therefore, necessarily have a zero rest mass. If they moved with a velocity v less than c, they would reverse their helicity for an observer moving faster than v in the same direction. Hence, their helicity would be dependent on the observer and could not be an intrinsic property.

With these helical neutrinos, the observed effects can indeed be explained (Figure 10). The emitted particles must take along some of the spin of the emitting nucleus, because it is known that the nucleus has a smaller spin after emission. This spin is a right-handed rotation when considered in the direction into which the spin points. Hence, the sense of rotation of the antineutrino must be the same as the one of the cobalt nucleus. Its direction of emission must then be such that a right-handed screw is formed. Hence, the antineutrino will be emitted only in one direction—namely, in the direction of the spin of the cobalt nucleus. It is a known property of the β-decay that the electrons are emitted mostly in the same direction as the neutrino. Thus, we get a preferred direction of emission for the electrons, as observed. In the mirror picture, the neutrino has

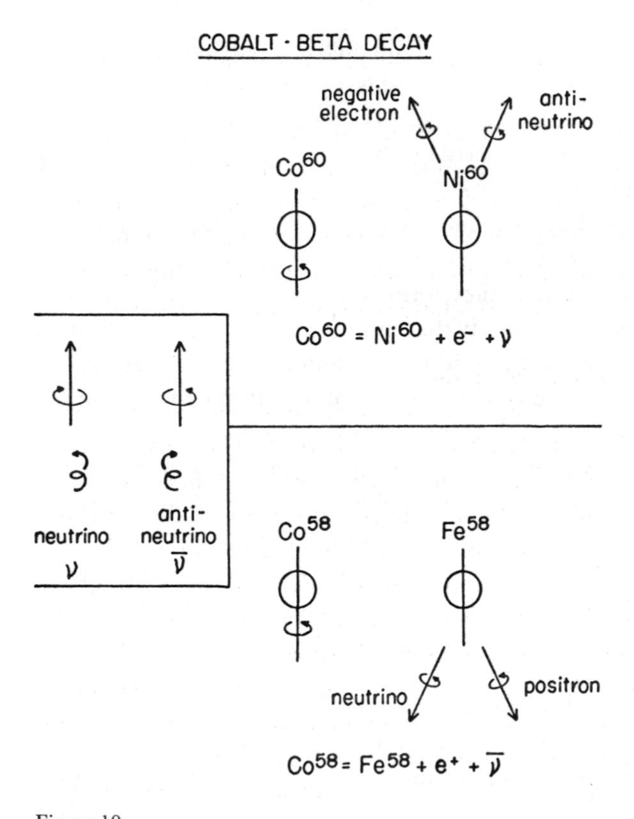

Figure 10.
The Lee-Yang-Landau explanation of the asymmetric beta decay of cobalt-60 and cobalt-58.

opposite helicity and therefore it will be emitted in the opposite direction.

A good support for this explanation is found in the experiment with cobalt-58, in which the emitted particles are a positron and a neutrino. If the hypothesis is correct, the preferred direction of the positrons must be opposite here to the preferred direction of the electrons in cobalt-60, for the neutrino has opposite helicity. In fact, that is just what the experiment has shown!

Experiments on Mesons

There is a second kind of experiments in which a similar violation of the parity law has been observed. These experiments have to do with some of the newly discovered short-lived particles, the mesons. The most important meson is the π-meson, which is probably the "quantum" of the nuclear force field. It is responsible for the binding forces in the nucleus. It occurs in three varieties, positive, negative, and neutral; it has a mass 265 times than of an electron, and it is known to have no intrinsic spin. When it is in free motion, the charged π-meson has a very short lifetime[7] of only 10^{-8} second and decays into a μ-meson and an antineutrino. The μ-meson is a particle very similar to an electron. It has a charge (positive or negative) and a spin of $\frac{1}{2}\hbar$ just like the electron, but its mass is 200 times larger. It too is unstable and decays after 10^{-6} second into an electron and a neutrino pair. This double decay chain

$$\pi \rightarrow \mu + \bar{v} \rightarrow e + 2\bar{v} + v$$

is a very interesting phenomenon and has been studied in detail.

Figure 11 shows a bubble chamber photograph of such processes, made recently by I. Pless, R. Williams, and coworkers. What one sees in such a picture are the charged particles only and not the neutrinos. One observes π-meson tracks coming

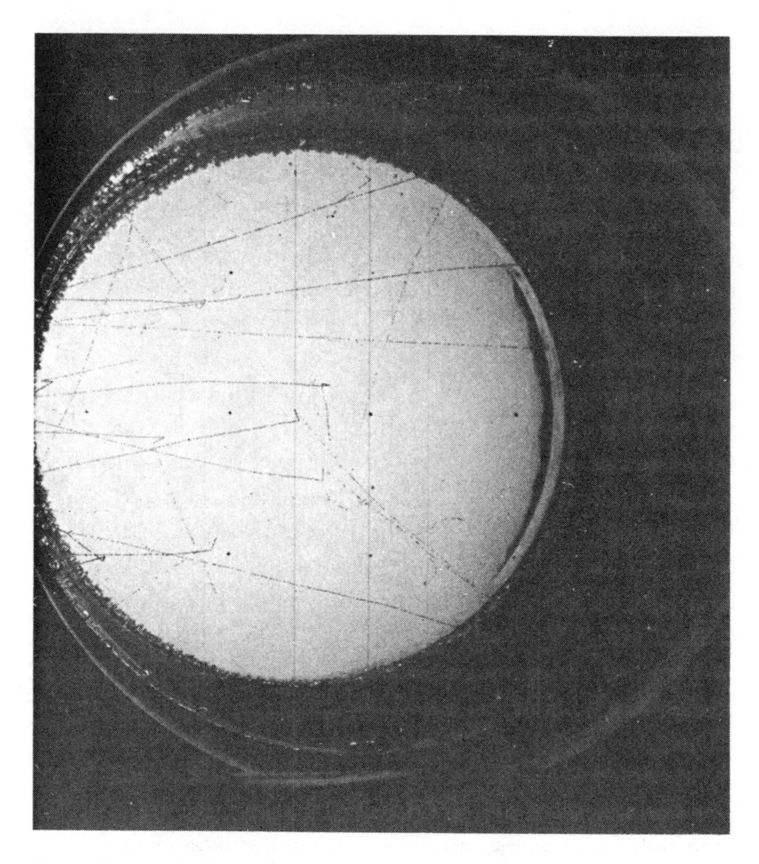

Figure 11.
Bubble chamber photograph of the π-μ-e decay chain (Pless and Williams). Dark tracks entering the chamber from the left are π-mesons. Short dark tracks at the ends of the π-meson tracks are μ-mesons produced in the decay of the π-mesons. The long light tracks are electrons produced in the decay of the μ-mesons. The electron tracks emerge in a predominantly backward direction relative to the direction of the μ-meson tracks.

from the left which end when the π-mesons come to rest. They then decay, and one sees a (short) μ-meson track emerge from the end-point of the μ-meson track. At the end of this track a third track emerges which is the track of the electron. The last track is longer again and is not very straight because the light-weight electron can easily be deviated from its path. A careful observer will find in Figure 11 that in five out of the six decay chains the electron is emitted "backward" in reference to the motion of the μ-meson. This effect has been established by more careful experiments, at Columbia University by Garwin, Lederman, and Weinrich, and at the University of Chicago by Friedman and Telegdi.

Why are the electrons emitted backward? Again, this is an example of the breakdown of the parity rule. When the μ-meson comes to a rest at the end of its short track, the only motion left is its rotation. How can a rotation determine a preferential direction of decay? Only by defining a preferential "handedness" or screw sense. This, of course, is a violation of the parity law, for the mirror image of the process would show the opposite preference.

Novelty of the Phenomenon

Let us now discuss two experiments which in all probability cannot actually be performed. A discussion of them is instructive, however, because it illustrates the essential novelty of the phenomenon.

We first return to the pivoted bar with which we began this discussion. Suppose it is made of cobalt-60, and suppose we rotate it about the axis AA'. (This example was suggested by E. M. Purcell.) As it rotates, the nuclear spins align themselves and the bar becomes very slightly magnetic. (This is the Barnett effect.) The electrons will then be emitted in a given direction; they will be absorbed in the bar, and one end will contain more energy than the other. (Actually, under normal conditions, this effect is so small that it cannot be observed at all.) Since the theory of relativity tells us that energy and mass are related, one end will be heavier than the other. Then, theoretically at least, the bar *will* tilt. Thus, a microscopic process (beta decay) which violates the principle of parity could lead in principle to a macroscopic observation of its violation.

An even more dramatic experiment has been suggested by J. R. Zacharias. Suppose a small round disk of aluminum is coated on the top with a thin film of cobalt-60 and suspended in a horizontal position by a thin wire attached to its center, as shown in Figure 12. The disk will begin to rotate! And, if the experiment is repeated, it will always rotate in the same direction! This can be understood from our previous discussion of beta decay, if we observe that the electrons which are emitted downward will be stopped in the aluminum, while those which are emitted upward will escape (the neutrinos escape in either case). One can think of the electrons which are stopped as transmitting their helicity to the block, which then begins to rotate. If the cobalt coating were on the lower side, the rotation would be in the other direction.

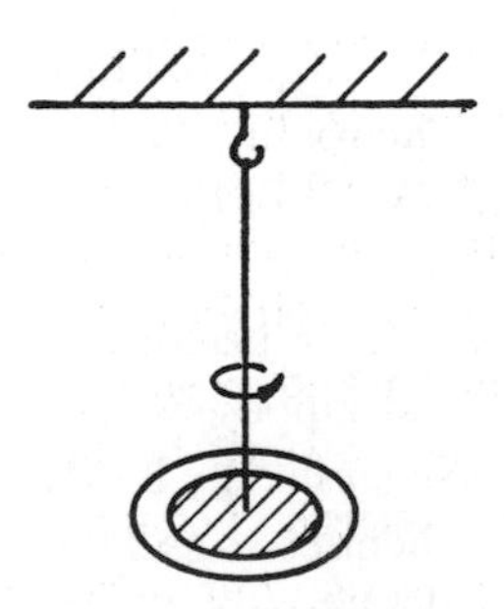

Figure 12.
Aluminum disk suspended by a thin wire. If the disk is coated on top with cobalt-60 it will spontaneously rotate as shown.

Antimatter

It is very suggestive to consider the violation of the principle of parity in connection with another somewhat better known asymmetry in our physical world. This is the asymmetry with respect to electric charge. The massive atomic nuclei are all positively charged, and the light electrons are negative. Physicists began to suspect that this asymmetry was only apparent after the discovery of the positive electron, the positron, in the early 1930s. It was shown that one can produce an electron pair, a negative and a positive electron, with light quanta of sufficient energy. The positron is in all respects the exact opposite of the negative electron; it is its so-called *antiparticle.* If a positron hits an ordinary electron, the two particles annihilate each other (the opposite process of pair creation), and their masses are transformed into light energy. The question of charge symmetry was completely cleared up after the discovery last year of the antiproton and the antineutron. The antiproton is a negative proton; it is antiparticle to the ordinary proton. It was produced with the very high energies now available from the large accelerators. The antineutron is the antiparticle of the ordinary neutron; it is just as neutral in respect to charge, of course, but it is opposite to the neutron in all respects. For example, it has the opposite magnetic moment, and it will annihilate into γ-rays or other forms of field energy with any neutron it meets, just as the negative proton will when it encounters a positive proton.

Hence, it seems that the charge asymmetry of matter is only apparent. One could also build up "antimatter," as it were, by using antiprotons and antineutrons for nuclei and positrons around them instead of electrons. Such antimatter would be the exact replica of our matter, with opposite charge: negative nuclei and positive electrons. It just so happens that our world is made of one type of matter. Some distant galaxies might be made of the other type.

We do not know much about the properties of antimatter, but it is highly plausible that there exists an interesting reciprocity in respect to the parity problem. We have mentioned that cobalt-58, which is a positron emitter, has shown the opposite helicity to the negatron emitter cobalt-60. Cobalt-58 emits neutrinos, which are the antiparticles to the antineutrinos emitted by cobalt-60. Hence, antiparticles seem to have the helicity opposite to that of the particles. Thus, it is most probable that "anticobalt-60" would emit its positrons in the opposite direction to cobalt-60. If this is so, the violation of the mirror symmetry appears in a new light: We argued before that the mirror image of cobalt-60 decay does not correspond to any possible process in nature. Now we see that this mirror image might be just the decay of "anticobalt-60"! By bringing together the two asymmetries in nature, the charge asymmetry and the mirror asymmetry, we might be at the threshold of the discovery of a new and higher *symmetry*, which Landau has called the combined parity principle. This principle says that the mirror image of any process in nature is also a possible process, but only if all charges are replaced by their opposite charges or if matter is replaced by antimatter. Since matter and antimatter are completely equivalent, the mirror symmetry of nature would be reestablished in a new and more interesting form.

We have seen in these developments how the increase in our knowledge of the properties of nature sometimes rocks the foundations of our understanding and forces us to a greater awareness of unsolved problems. The more the island of knowledge expands in the sea of ignorance, the larger its boundary to the unknown.

Remark Added by the Author, 1972.
In 1964, Christensen, Cronin, and Turlay discovered that there exist processes in which even the matter-antimatter symmetry is broken. They discovered what is called a *CP* violation. The *CP*

symmetry is the invariance of natural laws against charge conjugation and simultaneous parity conjugation. The former is the transition from matter to antimatter, the latter is the transition from a right-handed system to a left-handed one. *CP* symmetry is the one that was proposed in the last section of this article. It is the same as Landau's combined parity principle. The violation of this symmetry was shown in a somewhat intricate way by observing the decay modes of the two uncharged *K*-mesons, one of which, if *CP* is not violated, should decay only into two pions, the other into three. They found a slight amount of two-pion decay in the one that predominantly decays into three pions. A more direct observation of *CP* violation was made later when it was found that uncharged *K*-mesons do not decay with exactly equal rates in the following two processes:

$$K_0 \rightarrow \pi^+ + e^- + \nu, \qquad K_0 \rightarrow \pi^- + e^+ + \bar{\nu}.$$

One of these two processes is the "antiprocess" of the other and ought to be identical if matter and antimatter have the same properties. No right or left question enters into these processes.

The Role of Symmetry in Nuclear, Atomic, and Complex Structures

The concept of symmetry is used in everyday life. *Webster's International Dictionary* gives the following definition: "Correspondence in size, shape, and relative position of parts that are on opposite sides of a dividing line or median plane." *Webster's* also gives another definition: "Due or balanced proportions, beauty of form arising from such harmony"; adding that this meaning is "*now rare.*" We agree. The former definition is what comes first to one's mind, but it is incomplete. There are also translational (linear repetition), rotational (*n*-fold repetitions around a circle), or helical symmetries (a combination of the two) as well as centro-symmetry (repetition through a point), which go beyond *Webster's* definitions. Furthermore, symmetry in physics includes also correspondences which do not occur in space but which are valid in respect to some other dimension. There is a charge symmetry which implies that the antiworld with opposite charges—negative protons, positive electrons—is equivalent to ours; there is a mirror symmetry in respect to the time dimension, which says that elementary natural processes go one way in time equally well as in the opposite way. My remarks will be restricted to spatial symmetries only.

Within the framework of classical physics, spatial symmetry is more or less accidental; it is "arranged," or superimposed. Consider as an example the planetary system. According to Newton's laws, a symmetric circular orbit is highly improbable. The nice symmetric pattern of planetary orbits found on the letterheads of Atomic Energy Commissions are possible orbits (if they are well designed) but would occur with vanishing probability in a real system.

The irrelevance of symmetric shapes (shapes can be symmetric but need not be) is connected with a fundamental shortcoming of classical physics: The existence of atoms with their well-

Revision by the author of an article originally in *Nobel Symposium 11,* A. Engström and B. Strandberg, eds., John Wiley & Sons, New York, 1968.

defined characteristic properties and shapes is not explainable. Within the classical framework planetary systems should not exhibit any typical size or shape. It was quantum mechanics which solved this problem by bringing into physics a new "morphic" element, a characteristic pattern for a given state of a mechanical system: the Schrödinger wave function. These patterns reflect the fundamental symmetries of the mechanical constraints. For example, in the electric field of the atomic nucleus, the quantum states of electrons exhibit the symmetries of the spherical harmonic functions, a symmetry which is related to the spherical symmetry of the nuclear Coulomb field. The lower the energy of the state, the simpler the symmetry. The electron cloud around the nucleus at low energy is constrained to assume these shapes and no others. These patterns justly could be called the fundamental shapes of nature, *die Urformen der Natur*. (See Fig. 1 of the first article in Part 2.)

Another new concept, nonexistent in classical physics, is connected with this: the true identity of atoms when they are in the same quantum state. Classically, two atoms could be almost the same to any arbitrary degree, but it is infinitely improbable that their orbits are truly identical. In quantum physics, two hydrogen atoms in the ground state are truly identical. Quantum states are either identical or very different; there is no continuous transition; the characteristic symmetry is either the same or qualitatively different. Identity becomes a physically observable fact; it manifests itself, for example, in the spectrum of diatomic molecules when certain levels vanish for identical atom pairs.

Whenever atoms are formed from nuclei and electrons, the final products will be the same with the same properties, shapes, etc. There is only one pattern possible for electrons in the field of a nucleus with a given charge at its lowest energy. The uniqueness of this pattern guarantees that, whatever way an atom is assembled, it will always end up to be the same kind of atom. It is the basis of how atoms of the same species can be "replicated" from a "nutrient" of nuclei and electrons. We can go one step further down and regard the atomic nuclei as systems of protons and neutrons. Then a gas of protons and electrons is the nutrient for the replication of atomic nuclei and atoms (neutrons are produced by β-decay from protons), a replication which is not governed by a genetic code but by the basic laws of quantum mechanics.

If atoms of a given species are exposed to energy supplies which are high enough for their excitation—high temperature of a million degrees, for example—they loose their identical nature: Some will be found in one excited state, others in another. They lose their characteristic symmetries since many kinds of different symmetries are found in excited states. A mixture of many excited atomic states will not exhibit any particular symmetry.

Let us now turn to molecules. In many ways the characteristic shapes and properties of molecules are based on similar principles, but there are certain new points which come into play. At ordinary temperatures two molecules of the same kind are not necessarily identical in the quantum sense. This is because molecules are not found in their ground states at room temperature. Many of their rotational and deformational degrees of freedom are excited (there is thermal motion), and the probability of finding the two in exactly the same quantum state would be very small.

However, molecules have two kinds of excitation: One reflects the motion of the atoms within the molecule, the other represents the internal excitation of the atoms themselves. Only the first kind is excited at ordinary temperatures. Thus, the atoms within the molecule retain their quantum identity; the molecular structure, however, (the way the atoms stick together) can be suitably described in more classical terms. Hence the symmetry of molecules, in particular of macromolecules, is nearer to the everyday concept of symmetry. It is the symmetry of a structure composed of a number of identical parts, held together by some specific bonds. I would like to call it the "building block" symmetry, reminding you of that toy consisting of blocks with regular holes and sticks for joining them together. Two similar structures look alike, they have the same overall symmetry, but in principle, they can be distinguished because they differ certainly in some irrelevant respect. One may vibrate stronger at a specific bond, the other may rotate faster around another bond. Remember that two hydrogen atoms in their ground state are indistinguishable in principle.

Let us study a few interesting points regarding the "building block" symmetry. There are general geometrical rules about the symmetry of structures which are made of identical objects, regularly joined together. The simplest example is the linear structure: Whenever identical objects are joined, following the same rule each time for how the subsequent object is joined to the previous one, one always obtains a helical structure, provided that the parts do not get in each others' way. Of course, a straight or a circular arrangement are special cases of a helix. (Figure 1.) This simple but fundamental geometrical principle is rarely quoted in the literature in its generality, but it can be regarded as an explanation of the occurrence of helical structures in macromolecules. Clearly the helicity is assured, even if the elements are not exactly identical; only the spacial arrangement of the joint must be identical, and the deviations from identity must not be in the way of the helical structure, as it is in some proteins but not in the DNA or RNA chains.

Two-dimensional or three-dimensional assemblies of identical objects give rise to more than one type of structure. But here also there are a finite number of possibilities only: the 17 planar and the 230 three-dimensional crystal types. Structures which exhibit a "building block" symmetry may have imperfections in

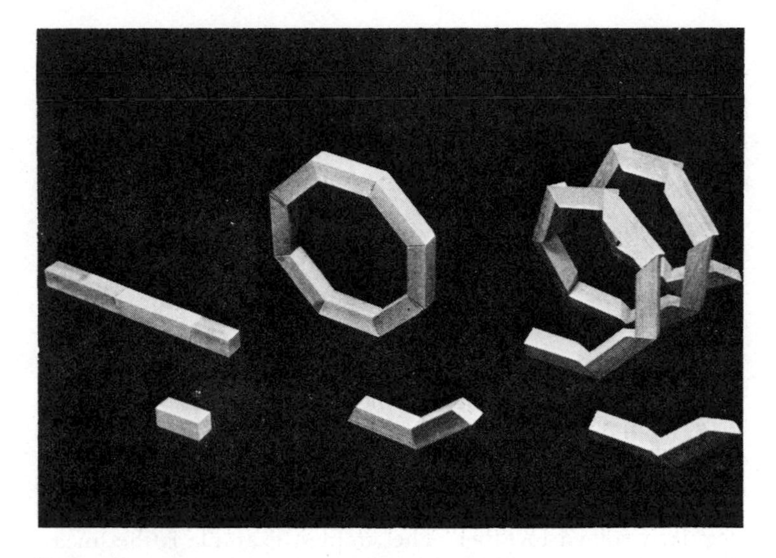

Figure 1.
Identical objects joined in a regular way produce a helix. The straight and the circular arrangements are degenerated helices.

their symmetries, in contrast to quantum states of atoms or simple molecules, where imperfections are unthinkable without any exterior perturbing field. The imperfections and irregularities of crystals are of interest in several respects. They form a structure of their own within the crystal; the strength of the material is determined to a great extent by the distribution of imperfections. They were introduced when the crystal was formed. Hence they give clues as to the history of the piece of material before us. They tell, as it were, the frozen story of how the crystal came about. They represent, therefore, what is interesting and specific about a chunk of crystal. The regular parts are universal and historyless; they tell nothing.

The imperfections in the symmetry of crystal structure tell the story of the past. The deviations from exact symmetry of macromolecules tell the story not only of the past but also of the future.

I am thinking here of the peculiar ordering of nucleotides in nucleic acids and of the amino acids in proteins. This is because there exists a mechanism in nature which, under certain conditions, replicates a macromolecule almost exactly, including its deviations from regularity. The moment such conditions are realized, the stage is set for a development which is called evolution. Some slight errors in replication give rise to new patterns in the order of units of the macromolecule. The patterns which are better able to reproduce themselves will become more numerous than the others. The biologists maintain that it is this mechanism that created the living world of today. The irregularities in the symmetry of macromolecules become assets instead of liabilities: We can speak of creative disorder. Here the dissymmetries tell not only the dead story of past happening; they are a link in an unending development towards more and more efficient and purposeful arrangements.

The symmetry of a structure which is capable of imperfections and disorder is fundamentally different from the kind of symmetry which we find in atomic quantum states. It is the building block type of symmetry. We may, but we do not need to arrange the blocks in a symmetric way. The nucleotides in a DNA molecule may be in complete symmetry: ATATAT . . . , but they may also follow each other in a different way.

A mechanism of molecular replication which replicates the molecular structure with all its patterns of dissymmetry must, in some ways, use the replicated structure as a template. This naturally excludes three-dimensional crystal structure as candidates for such processes. Two-dimensional structures are not excluded for template-like replication, but their mechanism is difficult to realize; in particular the process of separating the new sample from its template would need more energy than in the one-dimensional case; it is also less susceptible to catalyzers. This is why one-dimensional structures are the best candidates for carrying the genetic code as the basis of life. Since one-dimensional structures tend to be helical, we may have here a logical derivation of the helix as the basis of life and evolution.

What is the source of the macroscopic symmetries which we observe in living objects, the external bilateral symmetry of the animal body, the rotational symmetry of blossoms? I believe that these symmetries are also of the building block type, but the building blocks are not molecules but cells or cell complexes. The bilateral symmetry probably comes from the regularities set up in the first divisions of the fertilized egg cell.

I would like to mention here an interesting suggestion by C. S. Smith for an explanation of the frequently occurring five- and six-fold symmetries in flowers and perhaps also for the number five of fingers and toes. A statistical study of the shape of bubbles in froth has revealed that the polygons that are formed on each bubble by the lines of contact with adjacent bubbles are mostly pentagons or hexagons. In fact, the average number of corners of these polygons is 5.17. An assembly of cells should have a similar structure and it is suggestive to think that points of contact may give rise to special growth processes which may reflect the symmetry of the arrangement of these points.

There is no "building block" symmetry without identity among the blocks themselves. In molecular symmetries, this identity is based upon the fundamental symmetries and identities of atomic quantum states. But there is some further identity involved here which we have not yet discussed: The quantum states of atoms of the same species are identical because the electrons face exactly the same electric field of the nucleus. This sameness in turn is based upon the identity of all electrons and upon the identity among the nuclei of an atomic species. The latter identity is similar to the identity among atomic species; it comes from the fact that a given number of protons and neutrons, under the influence of the nuclear force, form always the same nuclear quantum state.

Hence, in the last analysis, the identity of atoms can be traced to the identity among themselves of protons, neutrons, and electrons. The reason for that identity is not yet known. The particle physicists are trying to solve this riddle. They want to find some fundamental and more abstract symmetry which allows elementary objects to exist only as nucleons and electrons. As usual in science, however, instead of solving this riddle, they found new riddles. When the structure of nucleons was explored with high-energy rays, new particles appeared, such as mesons and heavy electrons; the proton exhibits excited states of its own, as if it were a composite system itself. This is the exciting field with which my colleagues are struggling. I mention it here only in order to show how intimately the different fields of science are connected.

Molecular structure, the stability of macromolecules, the replication of a specific order, all these aspects essential for life are based upon identity and basic symmetry of atomic quantum states. These qualities, in turn, are based upon the specific properties of the elementary particles of matter. We cannot really claim an understanding of molecular processes before we know why the world consists of nucleons and electrons.

Part 4
General Essays

Part 4 contains articles about physics rather than on subjects of physics. The first, "Man and Nature," is an essay on the development of science in the last four centuries and the significance of this development as a unique phenomenon in the history of mankind and as a natural continuation of the evolution of life on earth

The second article, "Marie Curie and Modern Science," is based upon an address given at Marie Sklodowska Curie Centennial held in Warsaw, October, 1967. The development of nuclear physics is described and the significance of modern science as a supranational activity that serves to bridge political and national boundaries is stressed.

The third article, "Science and Ethics," is based upon a contribution to a symposium on science and ethics at the International Philosophic Congress in Vienna, 1968.

The fourth article, "The Significance of Science," is based upon several addresses that were given in 1970 and 1971 dealing with the role of science in the culture of today and with the many different interactions between science and society.

Man and Nature

The development of nuclear physics and its technical application to the harnessing of atomic energy for practical purposes is one of the most fascinating events in the recent history of human intellectual achievements. Twenty-five years ago practically nothing was known about the structure or the properties of the atomic nucleus, the massive center of the atom. In one decade of research so much information about this object was assembled that practical applications of some of the discovered features could be contemplated. By a dramatic coincidence of human history, nuclear fission was discovered just a few months before the outbreak of World War II. The possibility of creating a super-explosive was clearly indicated; the necessities of war intensified research and development to such an extent that the first production of nuclear energy was performed as early as 1942, when Enrico Fermi succeeded in producing a nuclear chain reaction in his first atomic reactor. After this, all work was concentrated upon the development of a new weapon of war. In the summer of 1945, only thirteen years after the first discovery of the neutron, two atomic bombs were exploded over Japanese cities, killing and maiming over half a million people.

So much terror and fear, so much emotion of guilt and pride are involved in this development and its effect upon the world that it is difficult to assess its significance at the present time. It is fortunate that now, after the tragic explosions, the emphasis has changed from military to peaceful applications of nuclear power. Here, as in so many other instances of exploitation of the forces of nature, the destructive use is the simplest one to devise but certainly not the only one possible. The use of nuclear energy for the production of industrial power and the

Revision by the author of an article originally in *Confluence* **5,** 1, 79, 1956. The author does not claim originality for the thoughts expressed in this essay. They represent a confluence in his mind of ideas by Julian Huxley, Gordon Childe, David Hawkins, and many others.

use of nuclear by-products in technology and medicine are just beginning. Nuclear power will not only supplement ordinary power sources but is bound to become the main source of energy, if human power requirements continue to expand at the present rate.

I

The remarkable thing about this development is its speed. The first insight into the structure of nuclei was gained in the late 1920s. The discovery of the neutron in 1932 led to the recognition that the atomic nucleus consists of neutrons and protons; it revealed the existence of strong forces between those particles which keep them tightly packed. These forces have been the chief interest of physicists ever since. They are different from any other force known so far in nature: They are not electric or magnetic, as are all chemical forces; they have nothing to do with gravity; they are much stronger than any other force between elementary particles. These nuclear forces are still being studied and their properties are far from being understood. However, the lack of comprehension of the fundamental forces has not prevented physicists from accumulating an enormous amount of data on the behavior and transmutations of atomic nuclei under bombardment by various particles.

In order to get an idea of the rapidity of the growth of nuclear science, we can compare it to the development of electricity. The discovery of the neutron in 1932 corresponds in some ways to the discovery of the electric current by Galvani and of the voltaic cell by Volta, both of which occurred around 1800. The discovery of fission in 1939 can be compared to Faraday's discovery of induction of currents by magnetic fields around 1830. The technical exploitation of electric power started with the construction of practical electric generators by Paccinotti and Siemens around 1860, sixty years after the discovery of the electric current; whereas the first useable atomic power pile was produced in 1954 in Russia, twenty-two years after the discovery of the neutron. This increased speed of development is all the more remarkable in view of the infinitely greater complexity of the nuclear problems compared with the electrical ones.

II

Let us take a more distant view of the development of science and technology in the last centuries. About 350 years ago, an increasing number of learned men directed their interest toward the observation of natural phenomena. Nature has always captured the interest of man; the mythologies of every age reflect the attempts to describe, explain, and influence the external world upon which man's existence and well-being depend. But only in the last three or four centuries have people hit upon the right methods and concepts for the study of nature. Knowledge about the natural world has accumulated in ever-increasing quantity and scope, and the exploitation of this knowledge has brought about a parallel growth in man's power to influence nature. The sudden development of our ability to cope with our environment is still so new to us that we cannot easily evaluate its significance. It is often depicted as purely materialistic, and of no great importance in the history of human thought. There can be no question, however, that it is a real objective phenomenon with all the characteristics of a fact of nature.

This is best seen by looking at it with the eyes of an extra human observer. Such a being would have noticed that the human population, more or less stationary till the Renaissance, suddenly and within a few hundred years increased fivefold. He would have observed that in many parts of the world the surface of the earth changed its character during this period, and that the change was obviously caused by human action. The most striking example was the continent of North America. Previously, it had been covered with tall dark green plants, but recently such plants had been cut down over vast stretches. The observer would have noted that the human animal had acquired a sort of metallic shell, which enabled him to move with a rapidity never previously known. The excretion of metals at a large rate was another new feature; these were first extracted underground and then spread in various forms over the surface of the earth in bulky structures, across rivers, and in thin double tracks across the country. These remarkable changes had taken place within a few generations; the observer from outside would be surprised about this speed, for all other living beings on the planet changed their structure and habits only over many thousands of generations.

The rapid development of the last few centuries is not a new phenomenon in the history of mankind. New social and technological patterns were established at the beginning of the Stone Age, the Bronze Age, and the Iron Age. This new kind of social evolution began a hundred thousand years ago with the appearance of the species man; before this, changes in biological structure and in social behavior went hand in hand. Millions of years had to pass before the external structure of an animal changed visibly. It took about sixty million years, for example, to develop the present horse from the eohippus, which, though similar, was a different animal distinct from the present horse. We have reason to believe that the social behavior pattern of animals did not change rapidly. It is more difficult, of course, to trace the social structure of a species than to trace the biological structure, but there is little evidence of a rapid evolution of social patterns of animals before the advent of man. The social organization of insects (bees, ants, or termites) does not seem to change during the span of time in which their bodily structure is fixed. In other words, the social structure of nonhuman animals seems to be just a part of their biological structure.

We might put down as an objective and "measurable" quality for distinguishing man from beast the fact that the changes in the organization of human behavior are extremely rapid compared with the changes in biological structure. The former changes take place within, say, one to ten generations depending upon the character of the period; the latter take place within about a hundred thousand generations: This constitutes a ratio of a factor of 10^5! The effect is by no means "marginal"; in fact, it is extremely strong and easily observable. In ten generations the behavior of any human race undergoes thorough qualitative changes; the difference in clothing, housing, and social behavior would be obvious to any "outside" observer. In order to obtain comparable structural changes in the human body, the time interval would need to be at least a hundred thousand times larger.

III

The evolution of living organisms is a phenomenon which has always attracted the interest of philosophers and scientists. How did it come about that life on this planet started several billion years ago with very primitive kinds of organisms and has developed more and more complicated structures? Science is very far from answering this question, but some of the more important factors in this process are beginning to take definite shape.

One of these factors is chemical reproduction. This is the ability of certain complex molecular structures to induce the formation of exact copies of themselves when they are placed in a solution containing the necessary chemicals. If one of these structures is put into such a solution, it initiates a continuously increasing growth of a population of similar structures until the material is exhausted. A well-known example of this process is the effect of placing a bacterium in a nutrient solution. The actual mechanism of this reproduction is not yet completely understood, but there is no question but that it is caused by the action of chemical forces not dissimilar from the ones which govern other understood chemical processes.

Further important factors involved in the evolutionary process are those of heredity, its stability and mutability. It seems that the reproduction of complex structures is governed by a central part in each structure which we call the nucleus. The situation is this: Any damage or change inflicted on the structure does not affect its ability to reproduce an undamaged new structure of the original design, except if the damage is inflicted on the nucleus. If the nucleus is harmed or changed, the structure loses its ability to reproduce, or it will reproduce a structure of *different* design whose nucleus shows the same changes that were inflicted on the first one. This new and different structure goes on reproducing and multiplying itself according to the new design. Changes in the nuclei which produce structures of different design are called mutations and are relatively rare events.

From these facts it is obvious that the nucleus contains a complete record of the design. Hence, it does not matter if other parts of the structure are damaged; the nucleus will provide for the reproduction according to the original design. If the nucleus itself undergoes change, however, then the record of design is changed and reproduction proceeds with a different design. We therefore say that the nucleus contains the hereditary material of the structure and determines its propagation.

On the basis of the two factors, reproduction and heredity, which occur with certain complex chemical structures, the development of life can be understood in its general lines according to Darwin's idea of natural selection: The process of mutation changes the design of the structures from time to time. Sometimes, though rarely, such changes lead to new structures which are better adapted to the environment, and these structures will become more prominent than the rest. Hence, we observe a constant evolution toward structures which are better equipped to find a suitable environment for multiplication.

It is not necessary to go into the details of the mechanisms which have evolved in this way since the beginning of life billions of years ago. The presently evolved animals are systems of enormous complexity adapted to their environment by many contraptions such as sense organs, nervous systems, a muscular structure for locomotion, a complicated system of chemical equilibrium. The development from the most primitive organism to the present living animal is in principle the effect of the interaction of the two factors, reproduction and mutation, of heredity. Clearly such a development is very slow. Doubts have been expressed by some people as to whether the time interval of several billion years would be enough for the accidental occurrence of all the mutations which were necessary for the evolution of the present forms of life. However, most biologists agree that the mechanism of evolution is based upon the natural selection of those combinations of mutations which give rise to the most favorably adapted organisms.

There is something truly remarkable about this mechanism: It provides a possibility for nature to improve itself, to "construct" more and more complicated structures, and to do this in a quite "natural" way without violating any of the fundamental laws of physics and chemistry and without having recourse to any pre-established plan. This process is all the more remarkable since we find in the nonorganic world more often than not the opposite trend: Complicated structures decay into less complicated ones, order goes toward disorder in an irreversible way. The build-up process is possible only because of the phenomenon of chemical reproduction. It automatically multiplies a more complicated structure if and when it is better adapted to its environment.[1]

A most important fact in the evolution of living structures is the lack of inheritance of acquired properties. A change in body structure inflicted upon an individual will never be inherited by its offspring. We can cut off the tails of all members of a group of animals and keep doing so to the offspring, but there will always be a tail on the newly born animals.

It is interesting to consider the reasons for and the limitations of this observation. A change inflicted on the body structure has no effect on the cell nuclei, and, in particular, leaves unchanged the nuclei of the germ cells which contain the blueprint of the new individuals. As long as the tail is planned for in the blueprint, it will develop in the offspring, irrespective of what has happened to the actual tails of the parents. There are certain limitations, however; in a large population of animals, there may from time to time be a few born without tails. These are the ones who by previous mutations and gene combinations possess a blueprint in their germ cells which does not provide for a tail. This, normally, is a rare case. But if some power from the outside continuously cuts off all tails, the ones born without tails will have a better chance of survival since they will not be exposed to the loss of blood caused by the cutting operation or to infections. Hence, after many generations, a tribe born without tails may evolve. This would simulate an inheritance of acquired properties, although actually it would only be the effect of a

[1]This trend toward complication in living structures is not in disagreement with the overall law of thermodynamics, which says that the total "entropy" (measure of disorder) must increase steadily. The increase of "order" in a living structure is always accompanied by a decrease of order in the sustaining physical environment. This balance is most important in the build-up of organic molecules in plants, which is done with the help of sunlight. For every molecule constructed, so much light energy is absorbed. This light energy was produced with a large loss of "order" in the solar material which emitted it.

selection of those strains which did not have a tail to begin with.

Perhaps the most interesting element in the functioning of living structures is the nervous system. This is a special combination of interlocking cells capable of transmitting stimuli from one part of the structure to the other. The formation of these systems makes it possible for the structure to acquire a "behavior." As a result they can then react upon changes in the environment in a much more varied fashion. Sense organs can be developed, and it is obvious that adaptation to the environment is greatly improved. With the formation of nervous systems comes the development of memories. The temporal sequence of certain stimuli produces one reaction; another sequence produces another reaction. It is possible to construct a nervous system where the type of reaction which follows a sequence of stimuli depends on other factors, such as the amount of food which this reaction provided the last time that it was performed. In this case the living structure "thinks," for it acts upon an impression from the outside only if it remembers that this action was favorable last time.

This is an example of a situation in which animal behavior can be crudely reproduced by a machine. Modern electronics has provided the means of constructing "nervous systems" of the simplest kind, and it is possible to build a model machine which exhibits rudimentary features of memory and premeditated action.

It is most plausible, therefore, that the behavior of an animal—its reaction to outside impressions and its ability to adjust to environment—is laid down by the design of its nervous system. This design, in turn, is contained in the chromosomes of each cell, and the tiny master plan is faithfully reproduced each time the cell divides itself in two. Hence the bodily structure of each individual animal and the basis of its social behavior is governed by some complicated secret code within the cell nuclei. When germ cells start the production of a new animal, this secret code determines the growth of the embryo, the development of limbs and organs, and the properties of the nervous system. Shape and behavior are predetermined in the chromosome. The animal develops, lives, acts, procreates, and dies without responsibility. It is not responsible for the structure of the chromosomes of its parents.

IV

In our tale of evolution we have reached the point where something new is beginning to develop, brought about by a simple increase in quantity of the cells which make up the nervous system. It happens often in the material world that an increase in quantity at a certain point gives rise to deep qualitative changes.

Let us look at the following example: An open vessel filled with water is enclosed in a room. When the temperature is below the boiling point, an equilibrium is established in which a certain number of water molecules per second evaporate from the surface and the same number per second return from the water vapor in the air and condense at the surface. The water in the vessel will remain seemingly undisturbed, in equilibrium with the (moist) air. When we raise the temperature but still keep it below the boiling point, there is only a quantitative change: There are just more molecules per second leaving and returning to the surface. However, if the number of molecules evaporating and condensing per second goes on increasing, a point will be reached where the returning ones can no longer keep up with the leaving ones. Then we will have reached the temperature at which the evaporation cannot be compensated for by condensation and the water is transformed completely into vapor: In other words, it boils away. To the onlooker it may seem that at the boiling temperature something special happens to the water. This is, in fact, not so; evaporation takes place also at lower temperatures. The decisive thing occurs in the relation of the water to the surrounding air. At the boiling point the air can no longer replenish the molecules lost by evaporation; hence, the evaporation, which was "harmless" to the water below the boiling point, "destroys" the water above this point.

We can observe a similar phenomenon in a solution of salt in water. If the concentration of salt is below the saturation point, the solution looks clear and no deposit is formed. Actually, however, the salt molecules do hit the walls of the vessel and form very tiny agglomerations, but the deposit is dissolved immediately. If the concentration of the solution is increased beyond the saturation point (e.g., by boiling off some water), the speed of formation of deposit surpasses the speed of redissolution; salt crystals begin to form in most beautiful patterns. Again, it would seem to the onlooker that at this point the solution has acquired a creative ability to give birth to special crystal structures. Actually, this is a quantitative question of the balance between deposit and dissolution.

Let us return now to the evolution of the nervous system in animals. We know that the nervous system enables animals to adapt themselves to their environment with the help of their sense organs and their memory. In fact, we know that animals "learn" from experience and that this learning capacity is an important factor in survival. However, a large part of animal behavior is based upon "instincts," is part of their biological inheritance.

When man evolved from the animal kingdom, something new must have happened which produced the avalanche effect that we observe in the development of man. We contend that this new element is based solely upon a quantitative difference in the nervous system. By an increase in size of this system, nature established a new type of evolution which has broken and will break all the rules established in the previous evolutionary periods.

How can these statements be substantiated? The elements of the new evolution are all present in the animal world: memory and learning and perhaps even the formulation of concepts and ideas. Only, as in a salt solution below the saturation point, they are yet too weak to be constructive. The attempts at learning in the animal world are mostly "dissolved" with the death of the individual.

There is no question but that animals have a language and that they have some means to transmit experience from one to the other. They may even be able to transmit experiences to the next generation. But these phenomena, especially the last one, are very weak and cannot overcome the dictation of animal behavior by biological instincts. When man evolved, the constant increase in the complexity of the brain and the nervous system reached a point at which the death of an individual no longer eradicated the gains which memory of experiences had established. The mechanism of this effect is quite obvious: The development of language and memory has enabled an adult individual to tell a younger one about his experiences and to impress him enough so that he acts as if he had had the experience himself. The workings of the brain have become complex enough to provide for vicarious experience and to enable man to pool the experiences of several individuals and eventually to *accumulate* experiences from generation to generation. This was made possible by the development of concepts, of mental constructions, of abstract ideas, and of many other methods of formulation and transmission of thought. It is most probable that the rudiments of all these means of thinking are present also in the minds of animals. The difference between man and animal is analogous to the boiling and saturation phenomena. Once the experiences collected by the species as a whole are more numerous than the experiences lost through the deaths of individuals, a new process begins which is the formation of a "tradition."

This is the point where evolution has overcome the barrier against inheritance of acquired properties. As long as parents cannot transmit their experience to their offspring, the behavior of each new generation is based exclusively upon biological inheritance; it is based upon what is inscribed in the master plan contained in the germ cells. The situation is not changed even if there is some transmission of experience from one generation to the next. Very probably there are instances among animals in which the young learn from their parents. As long as the sum of experience lost by death is larger than or as large as the sum transmitted to the next generation, there is no accumulation of experience. The behavior of each generation is essentially the same and is dictated by biologically inherited properties. However, if the transmission of experience between generations is large enough to cause an accumulation, the young will learn from the failures and successes of the elders, and newly acquired behavior patterns will be "inherited," not via the chromosomes but by word of mouth.

Once the critical number of nerve cells is reached and this stage of development attained, the further course is set and will develop at a constantly accelerating pace. Again the analogue of crystal formation in a saturated solution of salt is relevant. Crystal formation starts best from surfaces of other crystals. The first one formed has no such surfaces available. Hence, it takes a relatively long time to form the first small crystal. However, the next structures are formed at the surfaces of previously formed crystals. This makes for a rapid increase in the speed of formation. The greater the number of crystals formed, the greater are the opportunities for new formation. The same principle applies, then, to the formation of tradition. At the beginning, when mankind first acquired the possibility of developing it, the formation was very slow. Once started, however, it grew with increasing vigor and differentiation.

Tradition takes forms that are not always favorable to the

species. If, however, measures are found which are favorable, as, for example, agriculture, the exploitation of metals, etc., these measures initiate a new way of life within a few generations, and bring about that sudden change in behavior which is typically human.

Science is just one of these new measures or attitudes which grew from the accumulation of ideas and experiences. It took many generations to disentangle the vast number of observations, to separate apparent connections from real ones, to distinguish superstition from scientific fact. But once a systematic method for recognizing facts was found, the scientific revolution of the last 300 years started. There is no doubt that it constitutes an important step in the new kind of evolution which began with the formation of tradition. In our technological age, the inheritance of acquired properties is plainly visible. For example, we refer to the metallic shells in which the human species moves; they were acquired by one generation and inherited by those which followed.

So far, of course, it is only the pattern of behavior and thinking which is transmitted from one generation to the other. After all, the metallic shells are the product of systematic digging, burning, melting, and so on. The body structure is still reproduced in the old-fashioned animal way of propagation, and this leaves it unchanged for many generations. However, nobody can positively exclude a development like that portended in Aldous Huxley's *Brave New World*. It may become possible to change at will the chromosomes which determine the development of the species. Our knowledge of the mechanism of propagation is still very limited, but it grows dangerously fast, and human interference with the hereditary structure of germ cells is not completely out of sight. If this aim is achieved, the planned inheritance of desirable properties of the body will be within reach.

Even without having attained this ambitious aim, the new

evolution has left its mark on the planet and interferes everywhere, in an ever-increasing way, with the mechanism of the previous type of evolution. Man creates new races of animals by cross-breeding and purposeful selection. The natural evolution of the animal world is tampered with by human interference. When a new development in nature is discovered by man, it is channeled into some special direction. The times are over in which nature alone developed its own forms, slowly by trial and error, undisturbed over many generations.

V

The expulsion of man from paradise is a time-tested symbolic description of what happened when the new evolution set in. No longer do we rely upon cosmic rays to produce mutations and new forms and ways of life, with man as a happy onlooker. We now take it upon ourselves to develop nature and our own species. This is an arduous task full of pitfalls and responsibilities. We assumed this burden only a short time ago, and nobody should be surprised if we blunder now and then. After all, nature blundered in the previous evolution, when mammoths and dinosaurs acquired larger and larger dimensions until they were given up as dismal failures. We must proceed by trial and error, just as nature did. The pace of the new evolution by tradition, however, is very much faster than that of the old evolution by inheritance. Mistakes are punished immediately and cause tremendous suffering to the perpetrators and their offspring. We are responsible ourselves for what happens, and we cannot blame nature for it.

What is the background of this responsibility? The conditions of human life are the products of a behavior pattern based upon the accumulated experience and thinking of mankind. We compare our experiences with those of others and of previous generations and make decisions on how to act. In the time of the previous evolution, before man came on the scene, an act of decision taken by an animal was a purely individual matter, since the consequences of the decision were soon forgotten and did not influence the behavior of the rest of the world in any decisive way. In our present evolution, however, the consequences of a decision are not necessarily lost and forgotten. They are often remembered and incorporated in the accumulation of experiences which we call tradition. Hence, by making a decision, we add to human tradition, and our thinking may be contained in the future development, just as a successful mutation is contained in the hereditary material. This is where our responsibility comes in. We become responsible for the consequences of our thoughts, actions, and decisions when they contribute to human tradition.

It is quite obvious that the mechanism of the new evolution is still very mysterious and unpredictable. We cannot always foresee the effect of our actions upon human tradition. Although human beings create that tradition, it often seems to be developing according to its own laws, with an apparent inevitability independent of human control. For example, the progress of science and technology, once initiated by a few successful ideas, quickly assumed the character of an inevitable development. Conditions are created which make impossible any return to the status quo, and impasses are reached which can only be overcome by further evolution.

In this respect the new evolution is not different from the old. A striking example is the appearance of oxygen in the atmosphere. Before life in its present form developed on earth, the atmosphere did not contain any oxygen. Very probably it was this lack of oxygen which made possible the formation of the first complex organic molecule combinations. Presence of oxygen would have caused quick decay through oxidation and would have prevented the slow accumulation of organic molecules.

Plant life began with the evolution of molecular structures capable of transforming sunlight into chemical energy. This transformation was inevitably connected with oxygen production. Nature had run into an impasse: On the one hand, the transformation of sunlight into chemical energy was a very useful thing and helped living structures to withstand the rigors of life; on the other hand, the inevitable production of oxygen must have been harmful to the molecular structures, causing their decay. Plant life started out by "polluting" the air with oxygen. However, further evolution improved living structures sufficiently so that they could resist the oxygen pollution; and also, new species were created whose whole existence was based upon the presence of oxygen. The evolution of the animal world and eventually of mankind would have been impossible were it not for the initial air pollution with oxygen.

We do not know what the new evolution will eventually lead to. We have seen only the very beginning of this evolution, and we are at this moment (i.e., in the last 350 years) involved in a sudden outburst of a new variation in our tradition, the technological age. Who would dare, at this time, to say which features are good and which are bad? A pollution today may become a boon tomorrow and the blessings of today may cause the pollutions of tomorrow. The new evolution has given us the power of shaping the world. The only way open to us is to make use of this power in the best way we can. We struggle in the dark, for we do not know all the consequences of our actions. Jacob struggled with the angel all night until "the day breaketh. And he said: I will not let thee go except thou bless me."

Marie Curie and Modern Science

The century that has passed since the birth of Marie Sklodowska Curie has witnessed many events, changes, revolutions, and upheavals, which have left deep imprints on human society. Perhaps the most far-reaching change was the new insight that man acquired during this period into the basic workings of nature. It was in that century that the problems of the structure of matter were seriously attacked and, to a surprising extent, also solved. We know today, and we did not know a hundred years ago, many of the basic principles on which the behavior of matter is founded. We have acquired a far deeper understanding of what is going on in our environment; this knowledge also has enabled us to deal with our environment in a vastly more efficient way than before. It laid the ground for modern technology that has thoroughly revolutionized our way of life. The discoveries of Marie Sklodowska Curie initiated this development and are therefore in many ways symbolic of the new spirit of physical science.

One hundred years ago, physics was, to a large extent, a descriptive science. The question asked was "how," and not "why." Examples are the description of the motion of solids or liquids in mechanics and hydrodynamics, the description of the behavior of electric and magnetic fields by Faraday and Maxwell, the behavior of substances in chemistry. The chemists of that time described the reactions of atoms and molecules without explaining them.

During the lifetime of Marie Sklodowska Curie, a change occurred in the aims of physics: The question "Why?" was attacked in the study of material behavior. Physics changed from description to explanation. A great development took place which is characterized by the names of Planck, Einstein, Rutherford, Bohr, Sommerfeld, Franck, Schrödinger, Heisenberg,

Revision by the author of a lecture, "Introductory Talk for the Marie Sklodowska Curie Centennial," Warsaw, October 17, 1967.

Born, Pauli, and Dirac. It was the discovery of the quantum of action, of the nuclear atom, of quantized orbits, and, finally, of quantum mechanics, which is the way in which atomic behavior can be described and understood. The dynamics of the atom was discovered and cleared up, and with this insight all phenomena of the world of atoms and their aggregates, the facts of spectroscopy, of chemistry, of solid state, of material science fell into place and could be explained and understood as the effects of one fundamental interaction: of the electric force between atomic nuclei and electrons. Against the background of this development, let us look at Marie Sklodowska Curie's work.

We know what an important role radioactivity played, since it provided the necessary tool which enabled Rutherford to find the nuclear atom. But it was much more than this. Marie Sklodowska Curie's discovery was an anticipation of the next step that physics took more than thirty years later, a step that could only be taken after having gained insight into the workings of the atom: the search for the structure of the atomic nucleus. It introduced a new aspect into physics. When Marie and Pierre Curie isolated radium in that famous shed in the School of Physics in Paris, when they were awed by the uncanny gleam of that substance in the dark, they were looking at an extraterrestrial phenomenon, a phenomenon that goes beyond the ordinary atomic world of our environment. We know today that what the Curies saw was a remnant of the times when terrestrial matter was in a very different environment, in an exploding star. The natural radioactive substances are the last witnesses, the last embers still glowing, of the eventful times at which our elements were formed.

Thus Marie Sklodowska Curie's work initiated a third period in physics. It is the period which started in the 1930s, after quantum mechanics had explained the world of atoms. Then physics took a new step that we may call the *leap into the cosmos.*

It started with Marie Sklodowska Curie's discovery, but it really gathered momentum only after 1930, when systematic research was done to get at the inner structure of the atomic nucleus, to discover its composition, protons and neutrons, its dynamics, and its basic laws. A new force of nature revealed itself, the nuclear force between protons and neutrons, much stronger than other forces, with strange and apparently rather involved properties.

We should be aware that the processes and reactions revealed by this new branch of physics do not occur naturally on earth, except in those few processes found by Marie Sklodowska Curie. It is largely a man-made world, produced by our technical devices. But we have strong reasons to believe that this world plays a fundamental role in the universe; it is as essential for the interior of stars as atomic physics is for the surface of the earth. In terms of quantity of matter and energy, nuclear processes are relevant for a vastly larger part of the universe than atomic phenomena.

The end is not in sight yet of the leap into the cosmos by modern physics. Today our new accelerators penetrate into the subnuclear realm, into the internal structure of the nuclear constituents themselves. One discovers a new world of phenomena that lies beyond the world of nuclear processes; one finds excited protons, mesons, and heavy electrons; all this will lead to a deeper insight into the basic laws of matter. When it is better known what is going on within the proton and the neutron, we may understand better the nature of the nuclear force; it may be reduced to a simpler, more fundamental interaction, just as the chemical force was explained by the simpler and more fundamental electric interaction. Nuclear physics may become the "chemistry" of a new physics of elementary particles.

Today, we are only at the beginning of this great new development, which was initiated by Marie Sklodowska Curie's work. She and her collaborators and successors were able to deal with cosmic processes here on earth; they recreated such processes in our terrestrial environment. The investigation of these new phenomena was of tremendous importance for gaining new insight into the basic structure of matter and its behavior at high energy. But it has also another important aspect: The confrontation of nuclear processes with our terrestrial environment creates new effects and phenomena, which are of great scientific and practical interest. It must have been of special significance to Marie Sklodowska Curie that this confrontation did lead to practical consequences, even at the very beginning of this new science, when exposing living tissue to radioactive radiation led to such promising results. The treatment of cancer by the newly discovered rays should be a reminder to those people who believe that science should be interested only in the immediate environment of man and should not look out for new phenomena far removed from our daily experience.

But it happens rarely that application follows so quickly after discovery as it did in Marie Sklodowska Curie's work. Perhaps we were spoiled by this case and by the speed in which the application of nuclear energy followed the discovery of nuclear structure. In most instances a long time must pass before new discoveries are well understood and well in hand so that their interaction with our environment can be applied and made useful for other purposes. Imagine how long it would have taken to produce nuclear energy if there were less than two neutrons emitted per fission. A new and unusual field of research stands alone at the beginning and has no connections yet with other fields of interest. But whenever a completely new realm of phenomena is discovered, as it was in nuclear physics, as happens today in high-energy physics, there comes a time when the interaction of these phenomena with our environment will lead to unexpected effects and to a broader involvement with the rest of our scientific and technical interests. This is why it is shortsighted to judge the importance of a new field of research by its present state of application or relevance for other sciences. Science must proceed undirected, independent of any aims at application, in particular, in those fields where it penetrates into new and unexplored regions. It was the drive to find out where the radiation comes from in pitchblende that made Marie and Pierre Curie discover radium. They were led by the great curiosity of the true scientist about what goes on in nature. But Marie Sklodowska Curie's curiosity was paired with a deep concern for the human fate. She was the great discoverer in the shed of the School of Physics *and* the driver of an ambulance in the battlefields of the First World War.

Her life and her interest point to another human element in science: the supranational nature of science. I expressly use this term instead of "international." Science springs from a deeply human urge: to know and understand what happens around us. It is a language common to all human beings and therefore is above any nationality. Marie Sklodowska Curie was aware of this, right from the start, with her double nationality, Polish and French. In her later years she actively worked for the cause of international understanding by scientific collaboration. She saw that science is most potent in bridging the divisive forces of nationalism, racism, and different political systems. We may say without undue pride that the scientific community was to a large extent immune to prejudices of this kind and was most efficient in collaborating across geographical, racial, and political boundaries. Scientists knew perhaps better than others that only man counts and not the fact that he may be Jewish, Negro, or of any other origin. It is our task to pursue this tradition. We are facing today greater difficulties and dangers than ever before. Let me quote a statement by Marie Sklodowska Curie: "I believe that international work is a heavy task but that it is nevertheless indispensable; it must be pursued at the cost of many efforts and also with a real spirit of sacrifice. However imperfect it may be, the work of Geneva has a grandeur which deserves support." Of course, she had in mind the League of Nations which collapsed soon after. Perhaps there might be a consolation in the thought of how much she would have approved CERN, the new European Laboratory for high-energy research in Geneva, which is a truly supranational laboratory run by many nations, where scientists of the different countries work closely together and where any national origin vanishes when they enter the door. CERN is not the only one of such hopeful beginnings. There is a similar institution in Dubna north of Moscow run by a large group of nations.

But this is not enough; we must follow Marie Sklodowska Curie's great appeal and make our contacts between all centers of science stronger and more durable. There must be more interchange and common work between scientists from different parts of the world. The new tasks which science faces are so great that they require a common approach by all those who are participating. The costs are so high that they should not be wasted by lack of mutual help. Some steps were already taken in this direction: As an example, let me mention the close collaboration of European physicists at CERN, and the collaboration at the new high-energy center in Serpukhov, near Moscow where the Western European physicists join their skills with the Soviet physicists in order to exploit most efficiently the new 70-Gev proton accelerator. But this is not sufficient; why are there still so few so-called Western scientists working in so-called Eastern laboratories and so few from the East in the West? Why is the collaboration of American and Soviet scientists still subject to the vagaries of day-to-day politics? Why is there no Chinese physicist anywhere with us here?

The significance of scientific collaboration far exceeds the narrow aim of a more efficient prosecution of our scientific endeavors. It stresses a common bond between all human beings. Scientists, wherever they come from, adhere to a common way of thinking; they have a common system of values that guides their activities, at least within their own profession. New approaches in bringing nations together can perhaps be discussed with more ease within this community, some political misunderstandings can be cleared up, and dangerous tensions reduced. As an example, we recall that the agreement to stop the testing of nuclear bombs above ground stemmed in part from prior meetings between scientists.

We must keep the doors of our laboratories wide open and foster the spirit of supranationality and human contact, of which the world is so much in need. It is our duty to stick together, in spite of mounting tension and threatening war in the world of today. The present deterioration in the political world is a reason stronger than ever for closer scientific collaboration. The relations between scientists must remain beyond the tensions and the conflicts of the day, even if these conflicts are as serious and frustrating as they are today. The world community of scientists must remain undivided, whatever actions are taken, or whatever views are expressed in the societies in which they live. We need this unity as an example for collaboration and understanding, as an intellectual bridge between the divided parts of mankind, and as a spearhead toward a better world.

Science and Ethics

It is very difficult to talk about ethics in an objective and dispassionate way when we are witness today, as we were witness before, of human actions which are opposed to all ethical values. It seems so futile and irrelevant to talk about these values when they are violated all over the world.

I am not a philosopher but only a physicist who is deeply worried about the world. I cannot produce any profound or novel observations about ethics and science but I may present a few more or less obvious observations.

What is science and what is ethics? For the purpose of these remarks, I would like to restrict the term science to the natural sciences only. It is clear then, what is meant by this term. It is harder to say what is meant when one speaks of ethics. Let me follow Sir Karl Popper by using a negative characterization: It is connected with the prevention of violence of man against his fellow, and with the prevention of unnecessary suffering. But this touches only a part of it. The rest has much to do with the meaning of human existence and with the dignity of man, but I am unable to do more than point vaguely in that direction.

What is the connection between science and ethics? Most commonly one thinks of technology, which is the application of science. One observes that technology creates tools of violence which become more and more effective, dangerous, and easier to apply. Here science seems to run counter to ethical values. On the other hand, technology serves most efficiently in preventing human suffering first and foremost in medicine, which is the technology of life sciences. Furthermore, machines have replaced menial labor; fertilizers and modern agriculture technology are able to produce sufficient food for large segments of

Revision by the author of a contribution to the International Congress of Philosophy, Vienna, September 1968. Published in *Induction, Physics, and Ethics*, P. Weingartner and G. Zecha, eds., D. Reidel Publishing Co., Dordrecht, Holland.

the population. Altogether, thanks to technological achievements, it would be technically possible today to feed, clothe, and house the present population on earth without undue exploitation and suffering. Clearly we are far from this desirable state of affairs, and we are unable to cope with the ever increasing number of people, but the reasons for this shortcoming are no longer technological but social and psychological. Therefore, the problems of the improvement of the human condition must be attacked today on the political, sociological, and economic level. Science and technology have done their part already.

The two tendencies of technology toward destruction and construction are often considered as the center of the problem of science and ethics. It seems to me, however, that they have little to do with the problem. The issue is much older than science. From time immemorial man had to choose between helping and destroying. The question of ethics is not concerned with the ability to be helpful or destructive but with the ways and means to come to a decisive choice. Does science have any influence on this choice?

Here I see again two opposing complementary trends. In the past, human society has tied its ethical code to a supernatural system of thoughts which we commonly refer to as religion. It contains the idea of limitation of human capabilities, of the existence of superhuman forces which direct the lives of men toward a desired goal. There is a higher justice which punishes violations and rewards the fulfillment of the ethical code.

It can be maintained that science has shaken this system of thought. The growing success of science in explaining nature undermines the belief in the supernatural; the growing breadth of application of science has expanded human capabilities to an almost limitless extent: Human travel is expanding at a terrific rate and is no longer restricted to the surface of the earth; cosmic processes, such as nuclear fission and fusion, are made to work on earth. Almost everything seems possible, and almost everything is attempted. The respect, the fear, the awe of an imposed limit have disappeared and with them, the corresponding basis for an ethical code.

A parallel aspect of the influence of science on our thinking is often formulated as follows: Science represents nature as a mathematical formula and, therefore, has dehumanized our relations to the world. This is either expressed by stating that science considers everything in nature in a mechanistic way, to be nothing but machines or automatons, or by stating that science has dissolved matter into fields and energy; there is nothing absolute anymore, according to Einstein everything is relative and according to quantum mechanics all we see is but abstract vibrations. This view accuses science of regarding human feelings or emotions as irrelevant figments of the imagination.

A completely different view about the influence of science on our thinking is based on the following thoughts: Science endeavors to find the fundamental laws of nature which govern the world in which we live. It searches for the absolute, for the invariant in the flux of events. Science demonstrates the validity of natural laws to which the whole universe is subject. It finds and establishes insight and order where such was not found before. It creates a great edifice of ideas within which our natural environment becomes comprehensive and meaningful in its great development from a gaseous chaos to the living world. Relativity theory actually is a theory of the absolute and not of the relative, because it enables us to link and formulate the invariants in physics. Quantum mechanics does not dissolve matter into mere vibrations; it gives us a deeper insight into the nature of the material properties of atoms and molecules.

The development of science is one of the few evolutionary processes in human history. Contrary to a common belief, scientific theories are not overthrown; they are expanded, refined, and generalized. Einstein did not overthrow Newton's mechanics—the celestial satellites follow Newton's law—he expanded Newton's theory so that it applies also for very large velocities. Modern science is imbued with the same spirit as the science of Maxwell or Newton; it is only vastly more developed. This evolutionary trend comes from the collective nature of scientific work. The contribution of each individual scientist is based on the work of many others. No scientific achievement stands alone; it is always like a brick added to a single structure built by the scientific community over many generations. This is why that community has such strong internal bonds, which go beyond national and political boundaries. There exists not only a common terminology but also a common attitude among scientists which easily leads to creative and enduring intercourse between nations. This common attitude is somehow reflected by a certain state of mind within the scientific community which is conducive to a more positive outlook on the human predicament than in other social groups. It is a "happy breed of men" having a common task and believing—let me say, religiously—in the explicability of nature.

There is another aspect of science which Julian Huxley has emphasized. It recognizes the uniqueness of life on earth, which needed several billion years to develop. The loss of a single living creature means an irretrievable loss of a specific genetic combination. The pool of biological heritage which came to existence by the slow process of evolution would be irreplaceably lost if destroyed by a man-made catastrophe. It is our greatest responsibility to guard this pool and to continue its development, which today and tomorrow is in our hands for better or worse. From now on, we are responsible for the successful continuation of this extraordinary experiment which nature has started on earth. Certainly, these aspects of science have a deep influence on our thinking; they may lead to something like an ethical code, at least within the scientific community whose members are imbued with these ideas. These positive aspects of science also have some relevance for the nonscientific part of society, in spite of the fact that the nonscientist has little knowledge of the underlying ideas and concepts. There exists a general awareness of the far-reaching results and insights of science—we know something about the age of the earth and the universe, something about the origin of the elements, and about the evolutionary development of life, that matter is indestructible and, above all, there is an awareness among all people that nature works according to exact laws which exclude magic and demonstrate that man is not at the mercy of a capricious universe. These are edifying and stabilizing factors in the scientific picture of the world.

We have presented positive and negative aspects of the effect of science and technology on ethical values. The more powerful a complex of ideas is, the more influence it has upon thinking and action, the more ambivalent appears the outcome of this influence. In the history of mankind, ideas, situations, and opportunities have always been used and abused. The greatest ideas have led to the worst abuses. In this respect, our present situation is not too different from the situation in the past. But there is one element in our present predicament which seems to be different.

In order to be able to describe it, I must go further back in the development of mankind. Let me try to separate the human animal from other animals by the following distinction: In the animal world, the pattern of behavior, of customs and rituals, changes slowly at the same rate as the other biological features of the species. To a large extent, the behavior pattern is genetically fixed. Bees and ants behave as they have behaved since the species has evolved. The rate of change of behavior, therefore, is of the order of millions of years. In the human world, however, behavior patterns seem to change much faster; they are culture bound and not genetically fixed. The rate of change was perhaps of the order of several centuries in the last millennia, but this rate is accelerating. In the past, the rate has been very slow within one generation. Hence, it was possible to adjust to the cultural changes. The way of life of the parents was reasonably adequate to cope with the situation, since the behavior pattern did not change much from one generation to the next. The children could learn from the experience of their elders. It looks as if the changes today take place at an accelerated rate which is about to reach a new critical value; it changes so fast, that the behavior pattern is essentially altered within one generation. Once this rate of change is reached, we will face a qualitatively new situation which may be as different from the historic human situation as the latter one was from the animal world. I am not sure whether this analysis is correct. It may also be that there were times in the history of mankind when the rate of cultural change was as fast as today; these times may have been followed later by periods of stability. Whatever the historical truth may be, there is danger when the rate of change becomes too fast.

It is perhaps useful to analyze technological "progress" from this point of view. Would it imply that this progress is too fast and should be slowed down altogether? I do not believe so. In many instances, more progress is necessary to undo the damage past progress has done. One may be able to distinguish between progress which is stabilizing and progress which is destabilizing. Let me give two extreme examples: Progress in increasing the yield of food production, progress in developing efficient methods of birth control, are certainly stabilizing efforts. Further progress in air transportation, such as the supersonic transport, most probably is destabilizing. A great deal of technical, social, and psychological study would be necessary to establish in an objective and reliable way the stabilizing effect of technical developments. It is clear, however, that not everything is desirable which is technically feasible.

As Max Born has said: "Intellect distinguishes between the possible and the impossible; reason distinguishes between the sensible and the senseless. Even the possible can be senseless." Perhaps one can achieve an acceptable rate of change in our cultural patterns by a selection of fields in which progress should be supported or retarded.

The question may be raised whether progress of pure and basic science is a stabilizing or destabilizing factor. Quite apart from the fact that further progress of science may give us a freer choice between desirable and undesirable technological progress, the question is connected with the deeper problem of meaning and purpose of life, after the necessities of maintenance of life are established. I believe that the search for a deeper insight into what is around us is a fundamental human drive; it is one of these endeavors which give sense and dignity to our existence. To use our surplus energies and means for the exploration and understanding of nature, therefore, is an assertion of our human existence and our human pride. It is a stabilizing factor in the best sense, together with other creative activities in art and play. The collective character of scientific research makes large scientific enterprises suitable opportunities for the human urge to be active and productive on a large scale; it may help to channel aggressiveness into better directions than aggressive destruction.

Our aim is a society without violence of man against man, without unnecessary suffering and pain, without boredom and emptiness, a society in which a man can live a life of dignity and self-respect. Will we ever be able to reach these aims? I do not know, but we must live, think, and act under the definite assumption that we will.

The Significance of Science

And I gave my heart to seek and search out by wisdom concerning all things that are done under heaven. This sore task hath God given to the sons of man to be exercised herewith.

—Ecclesiastes I, 13

For in much wisdom is much grief and he that increaseth knowledge increaseth sorrow.

—Ecclesiastes I, 18

I. Three Positions

The development of science and technology during the last centuries has been very fast and overwhelming. All aspects of human society have been deeply influenced by it; the quality of life has been changed and often gravely disturbed. Today we have become very sensitive to the problems raised by this fast development, and we are faced with important questions regarding the role of science in society.

Science is under severe attack from some quarters; it is considered a panacea for the cure of all ills by others. I will sketch here three positions in regard to science that characterize some of the common attitudes towards this problem.

Position 1: Many branches of science have grown excessively during the recent decades; too large amounts of public support and too much scientific manpower are devoted to esoteric research in fields that have little to do with practical problems. Only such scientific research should be supported that promises reasonable payoff in terms of practical applications for industry, public welfare, medicine, or national defense. Science as the study of nature for its own sake is appreciated by only a few people and has very limited public value. Its support should be reduced to a much more modest scale.

Revision by the author of lectures given in 1970 and 1971. To be published later in *Science*, Spring 1972.

Position 2: Most of today's scientific research is detrimental to society, because it is the source of industrial innovations, most of which have led and will lead to further deterioration of our environment, to an inhuman computerized way of life destroying the social fabric of our society, to more dangerous and destructive applications in weaponry leading to wars of annihilation, and to further development of our society toward Orwell's world of 1984. At best, science is a waste of resources that should be devoted to some immediate, socially useful purpose.

Position 3: The methods and approaches used in the natural sciences and in technology—the so-called scientific method—have proved to be overwhelmingly successful in resolving problems, in elucidating situations, in explaining phenomena of the natural world, and in attaining well-defined aims. This method should be extended to all problems confronting humanity, because it promises to be equally successful in any area of human endeavor and human interest as it has been in the realm of natural science and technology.

These three positions are to a large extent mutually exclusive. They point in three almost orthogonal directions. In this essay I contend that each of these positions takes a narrow and one-sided view of the role of science in human society. Science is involved in man's thought and action in many different and often contradictory ways. Science must coexist with other forms of man's urges, feelings, and self-realizations. Science is based upon a very fundamental human urge, man's innate desire to know and understand the universe in which he lives and to gain insight into the driving forces governing the world around him. This urge is paired with the desire to improve the precarious conditions of human existence in an inhospitable natural environment and in competition with other human societies. Man desires to influence and to change the material and social conditions of life with the help of acquired knowledge and experience, which, in modern times, are mainly derived from science. As in all human situations, the urges and desires do not always lead to actions that serve the intended purposes, and the intended purposes are not always such that real benefits accrue for the people involved. These are the basic elements for our discussion of the role of science in human affairs. In the next three sections we will deal with some of the limitations of the three positions cited here. The subsequent sections are devoted to a more general appraisal of the situation.

II. Basic Science and Practical Applications

Let us return to Position 1, the excessive cost of basic science. It is based upon the supposition that most of research is unimportant and irrelevant if it is carried out without regard to practical applications. It is commonplace that technology and medicine owe an enormous debt to the study of nature for its own sake, that is to basic science. It is hardly necessary to mention here the many instances which prove that modern industry and modern care for the sick are based upon past results of basic science. Nor is basic science such an expensive luxury when its cost is compared with its services. The total cost of all basic science from Archimedes to the present day is probably near $30 billion,[1] less than 12 days' worth of production of the United States, whose gadgets and machines are to a large extent the product of earlier scientific achievement. The practical value of those parts of pure science which seemingly have no immediate connections with applications has been clearly brought out by H. B. G. Casimir, who collected a number of interesting examples of how decisive technical progress was made by scientists who did not work at all for a well-defined practical aim [1]:

[1]This figure is based upon an exponential increase of expenditure with a doubling time of 10 years, as it occurred during the last two decades, and a final yearly expenditure of 3×10^9 dollars. The starting time is irrelevant.

I have heard statements that the role of academic research in innovation is slight. It is about the most blatant piece of nonsense it has been my fortune to stumble upon.

Certainly, one might speculate idly whether transistors might have been discovered by people who had not been trained in and had not contributed to wave mechanics or the theory of electrons in solids. It so happened that inventors of transistors were versed in and contributed to the quantum theory of solids.

One might ask whether basic circuits in computers might have been found by people who wanted to build computers. As it happens, they were discovered in the thirties by physicists dealing with the counting of nuclear particles because they were interested in nuclear physics.

One might ask whether there would be nuclear power because people wanted new power sources or whether the urge to have new power would have led to the discovery of the nucleus. Perhaps—only it didn't happen that way, and there were the Curies and Rutherford and Fermi and a few others.

One might ask whether an electronic industry could exist without the previous discovery of electrons by people like Thomson and H. A. Lorentz. Again, it didn't happen that way.

One might ask even whether induction coils in motor cars might have been made by enterprises which wanted to make motor transport and whether then they would have stumbled on the laws of induction. But the laws of induction had been found by Faraday many decades before that.

Or whether, in an urge to provide better communication, one might have found electromagnetic waves. They weren't found that way. They were found by Hertz who emphasized the beauty of physics and who based his work on the theoretical considerations of Maxwell. I think there is hardly any example of twentieth century innovation which is not indebted in this way to basic scientific thought.

Some of these examples are evidence of the fact that experimentation and observation at the frontier of science requires technical means beyond the capabilities of technology. Therefore, the scientist in his search for new insights is forced to and often succeeds in extending the technological frontier. This is why a large number of technologically important inventions had their origin not in the desire to fulfill a certain practical aim but in the attempts to sharpen the tools for the penetration of the unknown.

The quoted examples are taken from past developments. It is frequently asserted that some branches of modern fundamental science are so *far removed* from the human environment that practical applications are most improbable. In particular, the physics of elementary particles and astronomy are considered to be in this category. These sciences deal with far-off objects; elementary particles in the modern sense are also "far off," because mesons and baryons appear only when matter is subject to extremely high energy, which is commonly not available on earth but probably occurs only in a few distant spots in the universe. The far-off feature of these sciences is also what makes them expensive. It costs much money to create conditions in our laboratory that may be realized only in some exploding galaxy. It costs much to build instruments for the study of the limit of the universe. The argument against these sciences is that they deal with subjects far removed from our human environment and that therefore they are of minor relevance.

What is human environment? Ten thousand years ago there were no metals in the human environment. Metals are rarely found in pure form in nature. But after man found out how to create them from ores, metals played an important role in our environment. The first piece of copper must have looked very esoteric and useless. In fact for a long time, man used it only for decorative purposes. But the confrontation of this new material with the human world brought about more and more interesting possibilities, resulting in a dominant role of metals in our environment. We have created a metallic environment. We even call historical ages by the typical metals used. For another example, electricity rarely appears in nature in its open form, only in lightning discharges and in frictional electrostatics. It was not an important part of the human environment. After long years of esoteric research into minute effects, it was possible to recognize the nature of electric phenomena and to find out what a dominant role they play in the atom. The introduction of these new phenomena into the human world created a completely new electric environment in which we live today, with 110-volt outlets in every wall. The most recent example lies in nuclear physics. In early days, prying into the problems of nuclear structure was considered a purely academic, esoteric activity, directed only at the advancement of knowledge about the innermost structure of matter. Rutherford said in 1933, "Anyone who expects a source of power from transformation of these atoms is talking moonshine." His conclusion was based on the same reasoning: The nuclear phenomena are too far removed from our human environment. True enough, apart from the rare cases of natural radioactivity, nuclear reactions must be artificially created at high cost with energetic particle beams. Most nuclear phenomena on earth are man made; they occur naturally only in the center of stars. But again, the introduction of these man-made phenomena into our human world has led to a large number of interactions: Artificial radioactivity has revolutionized many branches of medicine, biology, chemistry, and metallurgy; the process of fission is an ever increasing source of energy, for better or worse. Nuclear phenomena are now an important part of a new human environment.

These examples show the weakness of the argument that certain natural phenomena are too far removed to be relevant to the human environment. Natural laws are universal; in principle any natural process can be generated on earth under suitable conditions. Modern instruments create a cosmic environment in our laboratories when they produce processes that do not ordinarily take place in a terrestrial environment. Astronomy and particle physics deal with previously unknown and mostly unexplained phenomena. There is every possibility that some natural phenomena one day could also be reproduced on earth in some form or another and be applied—hopefully—in a reasonable way for some useful purpose. Today some special medical effects have already been found for pion beams, effects that cannot be brought about by any other means. E. Purcell [2] once said about the applicability of frontier fields such as particle physics: "In our ignorance, it would be presumptious to dismiss the possibility of useful application as it would be irresponsible to guarantee it."

One cannot divide the different branches of science into those that are and those that are not important for practical applications. The primary aim of science is not in application, it is in gaining insights into the causes and laws governing natural processes. But a better understanding of a natural process almost always leads to possibilities of influencing it, or of influencing other processes related to the one investigated. The further science develops, the more relations between seemingly unrelated processes are discovered. The study of the solar corona, for example, may lead to a better understanding of the behavior of highly ionized gases in magnetic fields, a topic of great technological importance. These interrelations between pure and applied science are part of the many-sided involvement of science in all aspects of human endeavor from the urge to know more about the environment to the urge to improve and dominate it.

III. Basic Science and Today's Problems

Position 2 is the expression of a widely held attitude that makes science bear the brunt of public reaction against the mounting difficulties of modern life. This is not the place to analyze in detail the predicaments of modern civilization, whose difficulties are related to the increased rate of technological expansion, a rate that today has seemingly reached a critical value both in time and in space. With regard to time, the changes in our way of life are now so rapid that marked differences are observable within one generation. This is a new and unsettling phenomenon for mankind; the experiences of the older generation are no longer as useful as they once were in coping with the problems of today. With regard to space, the effects of technology on our environment are no longer small; the parts of the earth's surface, of the water, of the air, which are changed by man or could be destroyed by man, are no longer negligible compared to those left untouched. These are unexpected and disturbing consequences with which we do not yet know how to deal.

Since technology, and in particular the increasing rate of technological change, is based largely upon science, it is not surprising that science is blamed for its difficulties. An obvious reaction to this situation would be to declare a moratorium on science; this would supposedly stop technological innovation and give us time to settle the problems that are already with us, instead of creating new ones. Recent cuts in scientific support reflect this attitude to some extent. We intend to refute, not the facts on which Position 2 is based, but the conclusions drawn from that position.

The call for a moratorium in science is based on its inexorable way of progressing further and further: One discovery leads to many others, and it seems impossible to prevent the application of new discoveries to unintentionally destructive purposes (such as polluting industries) and socially detrimental technologies.

Must we conclude, therefore, that it is detrimental and harmful to continue the search for further knowledge and understanding of the world in which we live? This search should be valuable under any circumstances, because knowing less about the world can hardly be better than knowing more. Ignorance is no value in itself; cruelty of man against his fellow man or thoughtless exploitation of man and nature existed long before the industrial revolution. To stop the growth of scientific knowledge would not prevent its abuses; it would deprive us not only of finding new means for dealing with them but also of an important source of cultural and philosophical insight. New scientific knowledge is neither good nor bad: No discovery has led only to destructive applications and not to constructive ones. New knowledge usually leads to a better way of predicting consequences and also sometimes to an ability to do something one could not do before. It is applied for good or for bad purposes, depending on the decision-making structure of the society; in this respect science and technology are not different from other human activities.

Today it is fashionable to emphasize the negative aspects of technological progress and to take the positive aspects for granted. One should remember, however, that medical science has doubled the average life span of man, has eliminated many diseases, and has abolished pain in many forms. It has provided the means of effective birth control. The so-called green revolution has created the potential to eliminate starvation among all presently living people. This is a scientific-technical achievement of momentous significance in the history of mankind, even though the actual situation is a far cry from what could be achieved. One should also remember the developments in transportation, construction, and power supply provided by modern technology and their great potentialities for improving the quality of life.

The trouble comes because technology hasn't achieved that purpose in too many instances. On the contrary, it has contributed to a definite deterioration of life. Medicine may have abolished pain, but modern weapons are producing wholesale pain and suffering. Medical progress has achieved a great measure of death control causing a population explosion; the available means of birth control are far from being effectively used.

The blessings of modern medicine are unevenly distributed: The lack of adequate medical care for the poor in some important countries causes mounting social tensions. The green revolution produces ten times more food than before, but the distribution is so uneven that starvation still prevails in many parts of the globe; furthermore, the massive use of fertilizers destroys many waters through eutrophication. Power production and the internal combustion engine as a means of transportation have polluted the atmosphere. Is it impossible to avoid harmful effects when we apply our knowledge of natural processes for practical purposes? It should not be so.

There are two distinct answers to these problems: the social and political aspect, and the technical aspect. In some instances the technical aspects do not pose any serious problems. The most important example is that of the use of technology for war or suppression. The only way to prevent the application of scientific results to the development of weapons is to reduce and prevent armed conflicts; certainly, this is a sociopolitical problem in which scientists and nonscientists should be equally interested, but it is not *per se* a problem of natural science. Other more benign examples are the problems of congested transportation, of city construction, and some, but not all, of the problems of pollution. In these cases we know what causes the trouble, and we know what measures can be taken to avoid it. But we don't know how to convince people to accept these measures. The problem is political and social: The methods of natural science cannot help except by pointing out as clearly as possible what the consequences of certain actions or inactions will be. It is beyond the scope of this essay to discuss whether one can resolve these political and social problems. We take the only possible attitude in this dilemma: We assume that there will be a solution at some time, in some form, to some of these problems.

But there are also many detrimental effects of technology of which the physical causes or the remedies are not known to a sufficient degree. Many detrimental effects of industrialization upon the environment belong in this category, such as carbon dioxide production, long-range influences on atmospheric currents and on climatic conditions, the influence of urbanization on health, the problem of better means of birth control, and many more. Here science has an enormous task in discovering, observing, and explaining unexplored phenomena, relations, and effects. The problems deal with our natural environment and therefore necessarily contain prime questions pertaining to natural science.

What role does basic science play in these efforts? One could conclude that the tasks are for applied science only and that research for its own sake, research that is not directed toward one of the specific problems, is not necessary. It may even be harmful since it takes away talented manpower and resources. *This is false.* The spirit of basic research is composed of the following elements: an interest in understanding nature; an urge to observe, to classify, and to follow up observed phenomena for the sake of the phenomena themselves; a drive to probe deeper into a subject by experimenting with nature, by using ingenuity to study phenomena under special and unusual conditions—all in order to find connections and dependencies, causes and effects, laws and principles. This attitude of basic research is necessary for the solution of today's pressing problems, because it leads people to search for causes and effects in a systematic way, independent of any ulterior aim. Many of today's troubles are caused by unforeseen consequences of human action on the environment, by interference with the natural cycle of events. The effects of accumulated technological developments are about to cover the entire earth's surface. We face a complicated network of physical, chemical, and biological causes and effects, many of them only partially understood. Much painstaking basic research will be required before these problems can be tackled efficiently. If technical solutions are introduced before the conditions are thoroughly understood, the situation may well worsen in the attempt to improve it.

Why is basic research needed for this kind of training? Why couldn't people be trained directly by putting them to work on socially pressing problems? Those who ask these questions compare the situation with teaching Greek and Latin to youngsters in order to give them experience in learning foreign languages. The comparison is fallacious. Polanyi [3] has expressed the reason most lucidly:

> The scientific method was devised precisely for the purpose of elucidating the nature of things under more carefully controlled conditions and by more rigorous criteria than are present in situations created by practical problems. These conditions and criteria can be discovered only by taking a purely scientific interest in the matter, which again can exist only in minds educated in the appreciation of scientific value. Such sensibility cannot be switched on at will for purposes alien to its inherent passion.

There are two sides to the argument: One concerns the analysis of a situation, and the other concerns the search for ways to improve it. The attitude engendered by pure science is most conducive to getting a clearer picture of the facts and the problems that may have to be faced in a given problem, such as air pollution, the population explosion, or the effects of technical innovations upon our environment. In basic science the search is for phenomena and connections in all possible directions, while in applied science the search is directed toward a specific goal. Furthermore, when new technical ideas are needed—and they will be needed—the attitude of basic science looks more toward innovative ideas and less toward the application of known devices, because the problems at the frontier are exactly those that cannot be solved with established methods. In basic research a pool is formed of young men and women who are accustomed to tackle unexplained phenomena and who are ready to find new ways to deal with them. They are trained to work under the most exacting conditions in open competition with the scientific world community. Instead of "environmentalists" we should train physicists, chemists, geologists, and biologists capable of dealing with the problems of environment.

Whenever large practical projects are carried out under emergency conditions, projects that appear immensely difficult or impossible, scientists from basic fields play a decisive role. In the past most of the examples have come from war-related projects, such as the development of radar or the atomic bomb. There is no doubt, however, that these experiences can be transferred to more constructive problems. In fact, many basic scientists have made important contributions toward a solution of the arms-control problem. Their activities initiated the discussions that led to the halt of bomb tests. Today basic scientists are deeply involved in environmental problems.

Two qualifications are in order: Today's problems certainly require the methods and results of natural science, but they cannot be solved by these methods alone. As mentioned earlier, the problems are to a great extent social and political, dealing with the behavior of man in complicated and rapidly evolving situations. These are aspects of human experience to which today's methods of natural science are not applicable. Seen within the framework of that science, these phenomena exhibit a degree of instability, a multidimensionality, for which our present scientific thinking is inadequate and to which such thinking must be applied with circumspection. There is a great temptation to transfer the methods that were so successful in natural science directly to social or political problems. But this is not possible for the most important problems. Adequate methods may be developed in the future. The social scientists are working hard at the task.

The second qualification concerns the need for scientists trained in basic science. We do not argue that *only* those trained in basic science to the exclusion of others can solve our problems. Far from it; a collaboration between all kinds of people is needed, basic and applied scientists, engineers, physicians, social scientists, psychologists, lawyers, and even politicians. The argument submits that people trained in basic science will play an important and irreplaceable role. They are necessary but not sufficient. But their necessity emphasizes the importance of keeping basic science activities alive.

Today to keep basic science vigorous is much harder than it was in the past; it would be harder even if the financial support were as generous as before. The reason is quite natural: The world situation has become so serious that many scientists or potential scientists find it difficult to worry about some unexplained natural phenomena or undiscovered law of nature when there are more immediate things to worry about. Some scientists feel that we are in an emergency situation and that we should stop basic science for the duration as we did during the Second World War. But the war lasted only four years for the United States, while the present crisis will endure for at least two decades. If we cripple basic science today, it will not be long before there will be no new generations of devoted young scientists for the tests that mankind must face in the future.

IV. The Limitations of Science

Another motivation for the antiscience attitude expressed by Position 2 is connected with the widespread critical view of science and its ways of thinking. In this view, science is considered as materialistic and inhuman, as trying to see everything in terms of numbers and thus excluding and denying the irrational and emotive approach to human experience. Value judgments, the distinction between good and evil, personal feelings, all supposedly have no place in science. Therefore, it is said, the one-sided development of the scientific approach has suppressed important and valuable parts of human experience; it has produced an alienated individual in a world dominated by science and technology in which everything is reduced to impersonal scientific facts.

These arguments are diametrically opposed to the views expressed in Position 3, which contends that the supposedly rational, inemotive approach of science is the only successful way to deal with human problems of all sorts. Much of today's trend against science is based upon the feeling that the scientific view neglects or is unable to take into account some of the most important experiences in human life.

This widely held belief seems to be in contradiction to the claim of "completeness" of science, which is the basis of Position 3. It is the claim that every experience—be it caused by a natural phenomenon or be it a social or psychic experience—is potentially amenable to scientific analysis and to scientific understanding. Of course, many experiences, in particular in the social and psychic realm, are far from being understood today by science, but it is claimed that there is no limit in principle to such scientific insights.

I believe that both the defenders and the attackers of this view could be correct, because we are facing here a typical "complementary" situation [4]. A system of description can be complete in the sense that there is no experience that does not have a logical place in it, but it still could leave out important aspects which, in principle, have no place within the system. The most famous example in physics is the complementarity between the classical description and the quantum properties of a mechanical system. The classical view of an atom, for example, is a little planetary system of electrons running around the nucleus in well-defined orbits. This view cannot be disproved by experiment; any attempt to observe accurately the position of an electron in the atom with suitable light beams or other devices would find the electron there as a real particle, but the attempt to observe it would have destroyed the subtle individuality of the quantum state which is so essential for the atomic properties. Classical physics is "complete" in the sense that it never could be proven false within its own framework of concepts, but it does not encompass the all-important quantum effects. There is a difference between "complete" and something we may call "all-encompassing."

The well-known claim of science for universal validity of its insights may also have its complementary aspects. There is a scientific way to understand every phenomenon, but this does not exclude the existence of human experience that remains outside science. Let us illustrate the situation by a simple example: How is a Beethoven sonata described in the realm of science? From the point of view of physics, it is a complicated quasi-periodic oscillation of air pressure; from the point of view of physiology it is a complicated sequence of nerve impulses. This is a complete description in scientific terms, but it does not contain the elements of the phenomenon that we consider most relevant. Even a psychological study in depth of what makes the listening to these tone sequences so exciting cannot do justice to the immediate and direct experience of the music.

Such complementary aspects are found in every human situation. There exist human experiences in the realms of emotion, art, ethics, and personal relations that are as "real" as any measurable experience of our five senses; surely the impact of these experiences is amenable to scientific analysis, but their significance and immediate relevance may get lost in such analysis, just as the quantum nature of the atom is lost when it is subjected to observation.

Today one is rather unaccustomed to think in those terms because of the rapid rise of science and the increasing success of the application of scientific ideas to the manipulation of our natural environment in order to make the process of living less strenuous. Whenever in the history of human thought one way of thinking has developed with force, other ways of thinking become unduly neglected and subjugated to an overriding philosophy claiming to encompass all human experience. The preponderance of religious thought in medieval Europe is an obvious example; the preponderance of scientific thought today is another. This situation has its root in a strong human desire for clear-cut, universally valid principles containing the answers to every question. However, the nature of most human problems is such that universally valid answers do not exist, because there is more than one aspect to each of these problems. In either of the two examples, great creative forces were released and great human suffering resulted from abuses, from exaggerations, and from the neglect of complementary ways of thinking.

These complementary aspects of human experience play an important role when science is applied to practical aims. Science and technology can provide the means and methods to ease the strain of physical labor, to prolong life, to grow more food, to reach the moon, or to move with supersonic velocity from one place to another. Science and technology are needed to predict the effects of such actions on the total environment. However, the decisions to act or not to act are based on judgments that are made outside the realm of science. They are mainly derived from two strong human motives: the desire to improve the conditions of life, and the drive for power and influence over other people. These urges can perhaps be scientifically explained by the evolution of the human race, but they must be regarded as a reality of human experience outside the scientific realm. Science cannot tell us which of the two urges is good or bad. Referring to the first rather than to the second urge, Archibald McLeish has put this idea into verse: "No equation can divine the quality of life, no instrument record, no computer conceive it—only bit by bit can feeling man lovingly retrieve it."

The true significance of science would become clearer if scientists and nonscientists were more aware of the existence of these aspects that are outside the realm of science. If this situation were better appreciated, the prejudice against science would lose much of its basis and the intrinsic value of our growing knowledge of natural phenomena would be much better recognized.

V. The Intrinsic Value of Science

Since the beginnings of culture, man has been curious about the world in which he lives and eager to explain it. The explanations have taken different forms, mythologic, religious, or magic, and they usually encompass all and everything from the beginning to the end. About 500 years ago man's curiosity took a special turn toward detailed experimentation with nature. It was the beginning of science as we know it today. Instead of reaching directly at the whole truth, at an explanation for the entire universe, its creation and present form, science tried to acquire partial truths in small measure, about some definable and reasonably separable group of phenomena. Science developed only when men began to restrain themselves not to ask general questions, such as: What is matter made of? How was the Universe created? What is the essence of life? They asked limited questions, such as: How does an object fall? How does water flow in a tube? etc. Instead of asking general questions and receiving limited answers, they asked limited questions and found general answers. It remains a great miracle that this process succeeded; the answerable questions became gradually more and more universal. As Einstein once said, "The most incomprehensible fact is the fact that nature is comprehensible."

Indeed, today one is able to give a reasonably definite answer to the question what matter is made of. One begins to understand the essence of life and the origin of the universe. Only a renunciation of immediate contact with the "one and absolute truth," only endless detours through the diversity of experience could allow the methods of science to become more penetrating and their insights to become more fundamental. It resulted in the recognition of universal principles such as gravitation, the wave nature of light, the conservation of energy, heat as a form of motion, the electric and magnetic fields, the existence of fundamental units of matter (atoms and molecules), the living cell, the Darwinian evolution. It reached its culmination in the twentieth century with the discovery of the connections between space and time by Einstein, the recognition of the electric nature of matter and of the principles of quantum mechanics, providing the answers of how nature manages to produce specific materials, qualities, shapes, colors, and structures, and finally the new insights into the nature of life provided by molecular biology. A framework has been created for a unified description and understanding of the natural world on a cosmic and microcosmic level, and its evolution from a disordered hydrogen cloud to the existence of life on our planet. This framework allows us to see fundamental connections between the properties of nuclei, atoms, molecules, living cells, and stars; it tells us in terms of a few constants of nature why matter in its different forms exhibits the qualities we observe. Scientific insight is not complete, it is still being developed, but its universal character and its success in disclosing the essential features of our natural world make it one of the great cultural creations of our era.

As part of our culture, science has much in common with the arts. New forms and ideas are created in order to express the relations of man to his environment. However the influence of science on society, on our lives and our thinking, is much greater today in the positive and negative sense; there have been times in the past in which the arts had a similar influence. Science is a unique product of our period.

Science differs from contemporary artistic creations by its collective character. A scientific achievement may be the result of the work of one individual, but its significance rests solely upon its role as a part of a single edifice erected by the collective effort of past and present generations of scientists. This effort was and is made by scientists all over the world; the character of the contributions do not reflect their national, racial, or geographic origin. Science is a truly universal human enterprise: The same questions are asked by all men involved in science; the same joy of insight is experienced when a new aspect of deeper coherence is found in the fabric of nature. The choices of problems, the directions of research at the frontier of fundamental science depend much less upon the economic, social, and political needs and pressures than most people assume; they are determined mainly by the instrumental possibilities of observation and by the internal logic of fundamental science itself. This is not so in applied science and technology, which obviously are much more —though not completely—subject to societal demands of all kinds. The rapid developments of applied electronics and acoustics during the Second World War were certainly determined by military needs. But there are exceptions on both sides: The progress in nuclear physics and in plasma physics—to a large extent a fundamental branch of physics—was certainly much accelerated by the possibilities of practical applications to power production by nuclear fission of fusion; the invention of the transistor—an example of applied physics—was not prompted by its practical potentialities.

It is difficult to distinguish clearly between fundamental and applied science, and any considerations of this kind can lead to dangerous oversimplifications. The success of basic research derives to a large extent from the close cooperation of basic and applied science. This close relation—often within the same scientist—provided tools of high quality, without which many fundamental discoveries could not have been made.

The scientific community is more international, or better, more supranational, than any other group because it transcends national and political differences. Personal contacts across borders are established easily between people working on similar problems; science has its own international language. The percentage of foreigners in scientific laboratories is probably greater than in any other human activity; there are some very successful international laboratories, among which CERN in Geneva stands out in the field of high-energy physics as a model for a future United States of Europe. The international ties of science have been helpful even in nonscientific affairs, for example, in the Pugwash Conferences, during which scientists initiated a number of actions directed toward a more unified world, such as ending atomic bomb tests in the atmosphere and beginning serious talks on arms control.

Science has a peculiar relation to the traditional and the revolutionary: It is both traditional and revolutionary at the same time. Newton's mechanics and the electrodynamics of Faraday and Maxwell are still valid and alive. Current calculations of satellite orbits and of radio waves are still based upon them. Revolutions, such as relativity and quantum theory, did not invalidate earlier ideas; they established unexpected limitations to the old ideas, which remain valid yet within the limitations. On the other hand, there is a strong trend in science toward the new and the different. Technological advances and novel ways of thinking are constantly introduced to change the manner of working and the method of approach. But scientific revolutions are extensions rather than replacements; apart from a few notable exceptions, old ideas are expanded and reinterpreted on a more universal basis. Old methods of observation are proved not wrong, but impractical and inaccurate.

In many ways, the attitude of mind in science is opposed to some of the negative and destructive trends in today's thinking. It means being involved in activities where there is real progress; deeper and deeper insights into the natural world are continually obtained. It engenders a feeling of participation at a unique collective enterprise, the construction and improvement of a vast intellectual edifice, one of the great creations of contemporary culture. There is little dispute among scientists regarding the general value scale of what is significant and as to the directions in which to proceed, although there are differences of opinion regarding the relative importance of different elements.

Scientific knowledge leads to an intimate relation between man and nature, to a closer contact with the phenomena derived from a deeper understanding. To know more about the laws and the fundamental processes on which the material world is based should lead to a deeper appreciation of nature in all its forms. It should show how almost every mineral structure and certainly every manifestation of life are unique and irreplaceable. Thus, science establishes an awareness of the importance and significance of each natural form, an awareness of how the universe, the atom, and the phenomena of life coexist and are all one. It is ecology in its widest interpretation.

There are still many fascinating problems and unanswered questions at all frontiers of science. We are not yet skillful enough to deal with complexity in nature. Even the structure of liquids is not well understood. No physicist would have predicted the existence of a liquid state from our present knowledge of atomic properties. The complexity of living matter presents far greater problems. In spite of the growing insight into the fundamental processes of reproduction and heredity, we still know very little about the development of organisms, about the functioning of the nervous system, and we know practically nothing about what goes on in the brain when we think or when we use the memory. The deeper we penetrate into the complexities of living organisms, into the structure of matter, or into the expanses of the universe, the closer we get to the essential problems of natural philosophy: How does a growing organism develop its complex structure? What is the significance of the particles and subparticles of which matter is composed? What is the origin of matter? What is the structure and the history of the universe at large?

The urge to find answers to such questions and to pursue the search for laws and meaning in the flow of events is the mainspring and the most important justification of science. These problems may have little to do with the practical needs of society, but they will always be in the center of interest because they deal with the where, whence, and what of material existence.

VI. Obligations of the Scientist

Does the actual science establishment correspond to the ideal picture of science as we have drawn it? It certainly does not appear so to many observers outside of the scientific community and even to some scientists. The human problems caused by the ever increasing development of a science-based technology are too close and too threatening; they overshadow the significance of fundamental science as a provider of deeper insights into nature. The scientist must face the issues raised by the influence of science on society; he must be aware of the social mechanisms that lead to specific uses and abuses of scientific results, and he must attempt to prevent the abuses and to increase the benefits of scientific discoveries. Sometimes he must be able to withstand the pressures of society toward participation in activities that he believes to be detrimental. This is not an easy task, because the problems are social in nature, and the motivations are often dictated by material profit and political power. It puts the scientist in the midst of social and political life and strife.

On the other side, the scientist also has an obligation to be the guardian, the contributor, and the advocate of scientific knowledge and insight. This great edifice of ideas must not be neglected during a time of crisis: It is a permanent human asset and important public resource. The scientist who today devotes his time to the solution of our social and environmental problems does an important job. But so does his colleague who goes on in the pursuit of basic science. We need basic science not only for the solution of practical problems but also to keep alive the spirit of this great human endeavor. If our students are no longer attracted by the sheer interest and excitement of the subject, we have been delinquent in our duty as teachers. We must make this world into a decent and livable world, but we also must create values and ideas for people to live and to strive for. Arts and sciences must not be neglected in times of crisis; on the contrary, more weight should be given to the creation of aims and values. And it is a great value to broaden the territory of the human mind by studying the world in which we live.

Much can and should be improved in the style and character of scientific teaching and research. The rapid increase of science activities during the fifties and sixties has left its mark upon scientists and science students. Some of the positive aspects have been adulterated; in many respects, science has become an organization for producing new results as fast as possible. Changes and new perspectives are in order. One of the most dangerous aspects in today's scientific life is overspecialization. There are several trends that lead to it. One is the increasing pace of research, which does not allow the researcher enough time to be interested in other fields not directly related to his own. He has enough trouble in trying to stay ahead of his numerous competitors in his own field and cannot devote much time to anything else. Another impetus has been the availability of research jobs in all fields; the young scientist therefore did not see the necessity of training himself in fields outside his speciality. Our educational system did not produce "physicists," it produced high-energy physicists, solid-state physicists, enzyme biochemists, and so forth. A typical symptom of this disease can be found in the manpower questionnaire that the National Science Foundation circulated among physicists, where one is asked to specify one's field, which is subdivided to the extreme. For example, there are divisions of this kind: elementary particles, hadron; elementary particles, lepton; solid state, magnetic properties; solid state, optical properties; . . . . And people try to find a job in exactly the subspecialty of their Ph.D. thesis. What a narrow view and what a boring life in the same subfield of physics forever! A physicist should be interested in all sides of physics and should welcome a change of field. Most of the positive aspects of science come from an awareness of its broad range, of its universal view. The same quantum theory governs elementary subparticles and phonons or excitons in a solid.

The teaching of science must return to the emphasis on the unity and universality of science and should become broader than the mere attempt to produce expert craftsmen in a specialized trade. Surely, we must train competent experts, but we also must bring together and show the connections between different fields of science. This task may be difficult because of heavy demands on the time and the intellectual capacities of those in modern research. But it is highly rewarding from any point of view. The teacher will get a deeper satisfaction from his work, and the student will enjoy his studies more; his knowledge will be broader, it will help him in his future work, and he will have a wider choice of jobs.

I. I. Rabi says so succinctly [5]:

> Science itself is badly in need of integration and unification. The tendency is more the other way . . . . Only the graduate student, poor beast of burden that he is, can be expected to know a little of each. As the number of physicists increases, each specialty becomes more self-sustaining and self-contained. Such Balkanization carries physics, and, indeed, every science further away from natural philosophy, which, intellectually, is the meaning and goal of science.

A broader understanding of science as a whole, beyond professional specialization, is a necessary condition for fostering the attitude toward nature that should be the basic philosophy of a scientist. It is that attitude of intimacy with the universe, with its richness and its uniqueness, the feeling of special responsibility toward nature here on earth, where we have power over it, constructive and destructive. The deeper understanding of nature as a whole leads to a duty on the part of the scientific community to be watchful and to warn against intentional and unintentional misuse of science and its applications.

Another destructive element within the science community is the low esteem in which clear and understandable presentation is held. This low esteem applies to all levels. The structure and language of a scientific publication is considered unimportant; content is all-important. So-called survey articles are understandable only to experts. The writing of scientific articles or books for nonscientists is considered a secondary occupation and, apart from a few notable exceptions, is left to science writers untrained in science. Something is wrong here. If one is deeply imbued with the importance of one's ideas, one should try to transmit them to one's fellows in the best possible terms.

In music, the interpretive artist is highly esteemed. An effective rendering of a Beethoven sonata is considered as a greater intellectual feat than the composing of a minor piece. Perhaps we can learn something here: A lucid and impressive presentation of some aspect of modern science is worth more than a piece of so-called original research of the type found in many Ph.D. theses, and it may require more maturity and inventiveness. Some students may derive more satisfaction from an interpretive thesis—as may some readers.

Furthermore, it is beneficial to the scientist to attempt seriously to explain his scientific work to a layman or even to a scientist in another field. Usually, if one cannot explain one's work to an outsider, one has not really understood it. More concerted and systematic effort toward presentation and popularization of science would be helpful in many respects; it would provide a potent antidote to overspecialization; it would bring out clearly what is significant in current research; and it would make science a more integral part of the culture of today.

Much more could and should be done to bring the fundamental ideas nearer to the intelligent layman. Popularization of science should be one of the prime duties of a scientist. The most important instrument for spreading the spirit of basic science is education. Young people should be more exposed than at present to the insights into nature that our age has revealed. There is more to it than the mere teaching of science: Scientific education must include active involvement in research. Students can absorb the spirit of science only if they face unsolved problems, participate in the process of analyzing facts, sift evidence, construct and test new approaches and ideas. Even at the lower levels, in elementary and high school, science activities should play an increasing role. Intelligent play involving simple natural phenomena fosters a deeper appreciation of our natural environment and transmits the joy of discovery. Margaret Mead [6] expressed it most impressively:

Any subject, no matter how abstract, how inanimate, how remote from the ordinary affairs of men, remains lively and growing if taught to young children who are themselves growing by leaps and bounds, hungering and thirsting after knowledge of the world around them. To children, an understanding of the world around them is as essential as the tender loving care that, during this century, has been so exclusively emphasized in discussions of early childhood education. The language of science will then become—for everyday use—a natural language, redundant, wide in scope, deeply rooted in many kinds of human experience and many levels of human abilities.

VII. Epilogue

Science is involved with society in many respects. There is a broad spectrum of relations, philosophic, social, and ethical, by which science influences and is influenced by society. The significance of science becomes evident in the numerous, often contradictory, ways in which it interacts with the affairs of men.

The philosophic significance of science is derived from the progressively deeper and more comprehensive insight into the workings of nature. The edifice of ideas that brought about this understanding of nature was erected during the last 300 years and is one of the most sophisticated systems of thought ever constructed by man. Its great power resides in the essential simplicity of the fundamental concepts. The infinitely complicated variety of phenomena seems to emerge from a few simple, though subtle laws of nature.

The social significance of science derives from the increasing ability to change the environment and the quality of life by applying the results of science. These changes have been both beneficial and detrimental, depending on the wisdom and the intentions of those who carried them out. They had a deep and lasting effect upon the social structure of society.

The ethical significance derives from the recognition that the evolution of life and men on earth is predicated upon a most precarious equilibrium of physical conditions on this planet. From this recognition follows a human responsibility to protect and continue the great experiment of nature that required several billions of years to get under way. Science emphasizes the unity of human beings in the urge to gain rational understanding of the workings of nature, and in the task of caring for our natural environment. It brings people together in their search for deeper insights, on a front that, to a great extent, remains uninfluenced by the political and social divisions.

Science claims universality. All phenomena and all human experiences are supposed to fit into the context of natural laws and are or will probably be scientifically described or explained. However, the scientific interpretation of human experiences does not always illuminate those aspects that are considered most relevant. These aspects include emotional experiences such as feelings and value judgements. They are decisive in the realm of human decision making. Whenever a choice is made between actions, whenever collective or personal decisions are taken, scientific reasoning can and should provide information about predictable consequences. The actual decision, however, remains outside of science, it represents a kind of reasoning that necessarily is complementary to scientific thought.

Science contains many activities of different aims and different character: the several basic sciences with all their variety of approach from cosmology to biology, and the numerous applied sciences that are spreading and involve ever more aspects of human concerns. Science is like a tree in which the basic sciences make up the trunk, the older ones at the base, the newer, more esoteric ones at the top where growth into new areas takes place. The branches represent the applied activities. The lower, larger ones correspond to the applied sciences that emerged from older basic sciences; the higher, smaller ones are the outgrowth of more recent basic research. The top of the trunk, the frontier of basic research, has not yet developed any branches. Applying this picture to the physical sciences, we would locate classical physics, electrodynamics, and thermal physics at the lowest part of the trunk with broad branches representing the vast applications of these disciplines. Higher up the trunk we would put atomic physics, with well-developed branches such as chemistry, materials science, electronics, and optics. Still higher we would find nuclear physics, with its younger branches symbolizing radioactivity, tracer methods, geology, and astrophysical applications. At the top, without branches, so far, we would locate modern particle physics and cosmology. There was a time, only sixty years ago, when atomic physics was the branchless top.

All parts and all aspects of science belong together. Science cannot develop unless it is pursued for the sake of pure knowledge and insight. It will not survive unless it is used intensely and wisely for the betterment of humanity and not as an instrument of domination by one group over another. Human existence depends upon compassion and curiosity. Curiosity without compassion is inhuman; compassion without curiosity is ineffectual.

References

[1] From a contribution by H. B. G. Casimir at the Symposium on Technology and World Trade, National Bureau of Standards, U.S. Department of Commerce, Wednesday, November 16, 1966.

[2] Quotation from an unpublished report to the Physics Survey Committee of the National Research Council, Washington, 1971.

[3] M. Polanyi, *Personal Knowledge,* University of Chicago Press, Chicago, Ill., 1958, p. 182.

[4] Similar views have been expressed by Thomas R. Blackburn, *Science* 172, 1971, p. 1003.

[5] I. I. Rabi, *Science The Center of Culture,* The World Publishing Co., New York, 1971, p. 92.

[6] Margaret Mead, "Closing the Gap Between Scientists and Others," *Daedalus* 88, 1959, p. 139.

Index